Pure Mathematics Unit 2
for CAPE®

Kenneth Baisden
Charles Cadogan
Sue Chandler
Mahadeo Deokinandan

Great Clarendon Street, Oxford, OX2 6DP, United Kingdom

Oxford University Press is a department of the University of Oxford.
It furthers the University's objective of excellence in research, scholarship,
and education by publishing worldwide. Oxford is a registered trade mark of
Oxford University Press in the UK and in certain other countries

First published by Nelson Thornes Ltd in 2013
This edition published by Oxford University Press in 2014

British Library Cataloguing in Publication Data
Data available

978-1-4085-2040-6

10 9 8 7 6

Printed and bound by CPI Group (UK) Ltd, Croydon, CR0 4YY

Acknowledgements

Cover photograph: Mark Lyndersay, Lyndersay Digital, Trinidad
www.lyndersaydigital.com
Page make-up and illustrations: TechSet Ltd, Gateshead

Thanks are due to Kenneth Baisden, Charles Cadogan, and Mahadeo Deokinandan
for their contributions in the development of this book.

Although we have made every effort to trace and contact all
copyright holders before publication this has not been possible in all
cases. If notified, the publisher will rectify any errors or omissions at
the earliest opportunity.

Links to third party websites are provided by Oxford in good faith
and for information only. Oxford disclaims any responsibility for
the materials contained in any third party website referenced in
this work.

Contents

Section 3 Counting, matrices and differential equations

Introduction

This Study Guide has been developed exclusively with the Caribbean Examinations Council (CXC®) to be used as an additional resource by candidates, both in and out of school, following the Caribbean Advanced Proficiency Examination (CAPE®) programme.

It has been prepared by a team with expertise in the CAPE® syllabus, teaching and examination. The contents are designed to support learning by providing tools to help you achieve your best in CAPE® Pure Mathematics and the features included make it easier for you to master the key concepts and requirements of the syllabus. *Do remember to refer to your syllabus for full guidance on the course requirements and examination format!*

Inside this Study Guide is an interactive CD which includes electronic activities to assist you in developing good examination techniques:

- **On Your Marks** activities provide sample examination-style short answer and essay type questions, with example candidate answers and feedback from an examiner to show where answers could be improved. These activities will build your understanding, skill level and confidence in answering examination questions.

- **Test Yourself** activities are specifically designed to provide experience of multiple-choice examination questions and helpful feedback will refer you to sections inside the study guide so that you can revise problem areas.

- **Answers** are included on the CD for exercises and practice questions, so that you can check your own work as you proceed.

This unique combination of focused syllabus content and interactive examination practice will provide you with invaluable support to help you reach your full potential in CAPE® Pure Mathematics.

1 Complex numbers and calculus 2

1.1 Complex numbers

Learning outcomes

- To define imaginary numbers
- To define complex numbers

You need to know

- How to solve a quadratic equation
- The relationship between the coefficients of a quadratic equation and the roots of the equation
- How to factorise a cubic expression

Imaginary numbers

When we try to solve the equation $x^2 + 1 = 0$, we get $x^2 = -1$ giving $x = \pm\sqrt{-1}$

Up to this point, we have left it at the statement that there is no real number whose square is -1, so the equation has no real roots. To work with equations whose roots are not real, we need to introduce another type of number.

If we introduce the symbol i to represent $\sqrt{-1}$, we can say that the equation $x^2 + 1 = 0$ has two roots, i and $-$i.

i is called an *imaginary number*, where $i = \sqrt{-1}$

It follows that $i^2 = (\sqrt{-1})^2 = -1$

This is consistent with what we know about the sum of the roots and the product of the roots of a quadratic equation: for the equation

$x^2 + 1 = 0$, $-\dfrac{b}{a} = 0$ and $\dfrac{c}{a} = 1$; and the sum of the roots is $i + (-i) = 0$

and the product of the roots is $i \times (-i) = -i^2 = -(-1) = 1$

Any negative number has two square roots, each of which is an imaginary number and can be expressed in terms of i.

For example, the square roots of -4 are $\pm\sqrt{-4} = \pm\sqrt{4} \times \sqrt{-1} = \pm 2i$
and the square roots of -49 are $\pm\sqrt{-49} = \pm\sqrt{49} \times \sqrt{-1} = \pm 7i$

Imaginary numbers can be added, subtracted, multiplied and divided.

For example, $\quad 2i + 7i = 9i$
$$i\sqrt{3} - i = i(\sqrt{3} - 1)$$
$$2i \times 7i = 14i^2 = 14 \times -1 = -14$$
$$10i \div 5i = 2$$

Powers of i can be simplified.

For example, $i^3 = (i^2) \times i = -i$, $i^4 = (i^2)^2 = (-1)^2 = 1$ and
$i^{-1} = \dfrac{1}{i} = \dfrac{i}{i^2} = \dfrac{i}{-1} = -i$

Complex numbers

Consider the quadratic equation $x^2 - 2x + 5 = 0$

The solution of this equation is
$$x = \frac{2 \pm \sqrt{4 - 20}}{2} = \frac{2 \pm \sqrt{-16}}{2} = \frac{2 \pm 4\sqrt{-1}}{2} = 1 \pm 2i$$

The two roots of the equation are therefore $1 + 2i$ and $1 - 2i$

These numbers are the sum of a real number and an imaginary number. Numbers of this form are called *complex numbers*.

A complex number is of the form $a + ib$ where a and b are real. a is called the real part and ib is called the imaginary part.

Like much mathematics that is devised to deal with a theoretical problem, complex numbers have many applications in real life. The main application of complex numbers is in electronics, where they are used to understand and analyse alternating signals. Mathematicians use i to denote $\sqrt{-1}$ but engineers use j to denote $\sqrt{-1}$ because i is used for current in electronics.

Conjugate complex numbers

The roots of the equation $x^2 - 2x + 5 = 0$ were found above to be $1 + 2i$ and $1 - 2i$

These two complex numbers are called ***conjugate complex numbers***.

Any two complex numbers of the form $a + ib$ and $a - ib$ are conjugate complex numbers and each is the conjugate of the other.

We use z to represent a complex number, so when $z = a + ib$, its conjugate, denoted by z^* (or by $\bar{z}$), is given by $z^* = a - ib$

The solution of the general quadratic equation $ax^2 + bx + c = 0$ is given by

$$x = \frac{-b \pm \sqrt{b^2 - 4ac}}{2a} = \frac{-b}{2a} \pm i\frac{\sqrt{4ac - b^2}}{2a}$$

Using $p = \frac{-b}{2a}$ and $q = \frac{\sqrt{4ac - b^2}}{2a}$, these roots can be expressed as $p \pm iq$

Therefore when a quadratic equation with real coefficients has complex roots, those roots are a pair of conjugate complex numbers.

We know, from Unit 1, that the left-hand side of a cubic equation with real coefficients can be factorised to give one linear factor and one quadratic factor. Therefore a cubic equation will always have one real root and when the other roots are not real, they will be a pair of conjugate complex numbers.

In fact, any polynomial equation with real coefficients can be expressed as a product of quadratic factors and possibly linear factors. Therefore any complex roots will be pairs of conjugate complex numbers.

Example

Find all the roots of the equation $(x^2 - 3x + 2)(x^2 + x + 2) = 0$

$(x^2 - 3x + 2)(x^2 + x + 2) = (x - 2)(x - 1)(x^2 + x + 2)$

The roots of $(x^2 + x + 2) = 0$ are $\dfrac{-1 \pm \sqrt{1 - 8}}{2} = -\dfrac{1}{2} \pm i\dfrac{\sqrt{7}}{2}$

$\therefore$ the roots of $(x^2 - 3x + 2)(x^2 + x + 2) = 0$ are $1, 2, -\dfrac{1}{2} - i\dfrac{\sqrt{7}}{2}, -\dfrac{1}{2} + i\dfrac{\sqrt{7}}{2}$

Exercise 1.1

1 Simplify:

(a) i^5

(b) i^{-5}

(c) i^{2n}

(d) i^{4n}

(e) i^{8n+1}

(f) $5i \times 2i$

(g) $8i \div 2i$

2 Find all the roots of each equation.

(a) $x^2 + 5x + 8 = 0$

(b) $x^3 + 2x - 3 = 0$

Learning outcomes

- To add, subtract, multiply and divide complex numbers

You need to know

- The form of a complex number
- The meaning of conjugate complex numbers
- The relationship between the coefficients of a quadratic equation and the sum and product of its roots

Equality of complex numbers

Two complex numbers, $a + ib$ and $c + id$ are equal if and only if the real parts are equal and the imaginary parts are equal,

$$\text{i.e.} \quad a + ib = c + id \Leftrightarrow a = c \text{ and } b = d$$

Addition and subtraction of complex numbers

The real parts and the imaginary parts are added and subtracted separately.

For example, $\quad (2 + 3i) + (5 - 2i) = (2 + 5) + (3i - 2i)$

$$= 7 + i$$

and $\quad (2 + 3i) - (5 - 2i) = (2 - 5) + (3i - (-2i))$

$$= -3 + 5i$$

Multiplication of complex numbers

Two complex numbers are multiplied together in the same way that we expand $(a + b)(c + d)$

For example, $\quad (2 + 3i)(5 - 2i) = 10 - 4i + 15i - 6(i^2)$

$$= 10 + 11i + 6$$

$$= 16 + 11i$$

and $\quad (2 + 3i)(2 - 3i) = 4 - 6i + 6i - 9(i^2)$

$$= 4 + 9$$

$$= 13$$

The fact that $(2 + 3i)(2 - 3i) = 13$ is a particular case of the fact that the product of a pair of conjugate complex numbers is a real number.

This is because $\quad (a + ib)(a - ib) = a^2 - (ib)^2 = a^2 + b^2$

Division of complex numbers

We can divide one complex number by another complex number by multiplying the numerator and the denominator by the conjugate of the denominator. This gives a real denominator.

For example, $\quad \dfrac{2 + 3i}{5 - 2i} = \dfrac{(2 + 3i)(5 + 2i)}{(5 - 2i)(5 + 2i)}$

$$= \dfrac{10 + 19i + 6(i)^2}{25 + 4}$$

$$= \dfrac{4}{29} + \dfrac{19}{29}i$$

Example

Find the values of a and b where $z = a + ib$ such that $2z + 3z^\star = 5 - 2i$

$2z + 3z^\star = 2(a + ib) + 3(a - ib) = 5a - ib$

$\therefore \quad 2z + 3z^\star = 5 - 2i \Leftrightarrow 5a - ib = 5 - 2i$

Equating real and imaginary parts gives

$$5a = 5 \Rightarrow a = 1$$

and $\quad -b = -2 \Rightarrow b = 2$

Example

One root of a quadratic equation with real coefficients is $3 - i$.
Find the equation.

Let the equation be $ax^2 + bx + c = 0$

If one root is $3 - i$, the other root is its conjugate, $3 + i$

The sum of the roots is $(3 - i) + (3 + i) = 6$

$\therefore \quad -\dfrac{b}{a} = 6$

The product of the roots is $(3 - i)(3 + i) = 9 + 1 = 10$

$\therefore \quad \dfrac{c}{a} = 10$

The equation is $x^2 - 6x + 10 = 0$

Example

Find the values of x and y for which $(3 - 2i)(x + iy) = 16 + 11i$

$(3 - 2i)(x + iy) = 3x + 2y - 2ix + 3iy$

$\therefore \quad (3 - 2i)(x + iy) = 16 + 11i \Leftrightarrow 3x + 2y - 2ix + 3iy = 16 + 11i$

$\qquad\qquad\qquad\qquad\qquad \Leftrightarrow 3x + 2y = 16 \ [1] \quad$ and $\quad -2x + 3y = 11 \ [2]$

Equating real and imaginary parts

Solving [1] and [2] simultaneously gives

$2 \times [1] + 3 \times [2] \qquad\qquad 13y = 65$

$\qquad\qquad\qquad\qquad\qquad \Rightarrow y = 5$

from [1] $\qquad\qquad\qquad\qquad x = 2$

Exercise 1.2

1 Find, in the form $a + ib$

  (a) $(2 - 4i)(-1 + 2i)$

  (b) $\dfrac{2 - i}{3 + i}$

  (c) $\dfrac{3}{2 - i} + \dfrac{4 + 3i}{1 + 2i}$

2 One root of a quadratic equation with real coefficients is $3 + 5i$. Find the equation.

3 Find the values of x and y for which

  (a) $(x + iy)^2 = 5 - 12i$

  (b) $\dfrac{x + iy}{3 - 4i} = 2 + i$

4 Find the values of a and b where $z = a + ib$ such that $z(2 + i) + 2z^\star = 4 - 5i$

- To find the square roots of a complex number

- How to multiply complex numbers
- The relationship between the roots and coefficients of a quadratic equation

The square roots of a complex number

If $x + iy$ is a square root of the complex number $a + ib$, then $(x + iy)^2 = a + ib$

Expanding the left-hand side gives $x^2 - y^2 + 2ixy = a + ib$

Equating real and imaginary parts gives a pair of simultaneous equations which we can solve to find values for x and y.

The equations are quadratic, so there will be two values for x and y.

Therefore a complex number has two square roots.

Example

Find the square roots of $3 - 4i$

If $x + iy$ is a square root of $3 - 4i$, then $(x + iy)^2 = 3 - 4i$

$\Rightarrow \qquad\qquad\qquad\qquad x^2 - y^2 + 2ixy = 3 - 4i$

Equating real and imaginary parts gives

$$x^2 - y^2 = 3 \qquad\qquad [1]$$
$$2xy = -4 \qquad\qquad [2]$$

$[2] \Rightarrow \qquad\qquad y = -\dfrac{2}{x} \qquad\qquad [3]$

$[3]$ in $[1] \Rightarrow \quad x^2 - \dfrac{4}{x^2} = 3$

$\Rightarrow \qquad x^4 - 3x^2 - 4 = 0$

$\Rightarrow \quad (x^2 - 4)(x^2 + 1) = 0$

x is a real number, so $x^2 = -1$ does not give a valid value for x

$\therefore \qquad x^2 = 4 \Rightarrow x = 2 \text{ or } -2$

from $[3] \qquad\qquad y = -1 \text{ or } 1$

$\therefore$ the square roots of $3 - 4i$ are $2 - i$ and $-2 + i$

Note that this example shows that if z is one square root of a complex number, $-z$ is the other square root.

Exercise 1.3a

1   Find the square roots of

    (a) $-2i$ (b) $-8 - 6i$ (c) $-1 + 2i\sqrt{2}$

2   $4 + 3i$ is one square root of $a + ib$

    (a) Find the values of a and b.

    (b) Find the other square root.

Quadratic equations with complex coefficients

A quadratic equation such as $z^2 + (6 + i)z + 10 = 0$ can be solved using the formula, which leads to finding the square root of a complex number.

For example, to solve $\quad z^2 + (6 + i)z + 10 = 0$

$$\Rightarrow \quad z = \frac{-(6 + i) \pm \sqrt{(6 + i)^2 - 40}}{2}$$

$$\Rightarrow \quad z = \frac{-6 - i \pm \sqrt{-5 + 12i}}{2}$$

If $a + ib = \sqrt{-5 + 12i}$, then squaring both sides and equating real and imaginary parts gives

$$a^2 - b^2 = -5 \qquad [1]$$

and $\qquad\qquad 2ab = 12 \qquad [2]$

From [2] $\qquad\qquad b = \dfrac{6}{a} \qquad [3]$

[3] in [1] $\Rightarrow \quad a^2 - \dfrac{36}{a^2} = -5$

$$\Rightarrow \qquad a^4 + 5a^2 - 36 = 0$$

$$\Rightarrow \qquad (a^2 + 9)(a^2 - 4) = 0$$

$\therefore \quad a = 2$ and $b = 3$ or $a = -2$ and $b = -3$

i.e. $\quad \sqrt{-5 + 12i} = 2 + 3i$

Hence $\quad z = \dfrac{-6 - i \pm (2 + 3i)}{2}$

$$\Rightarrow \quad z = -2 + i \text{ or } z = -4 - 2i$$

Notice that the roots of this equation are not a pair of conjugate complex numbers. Roots are only pairs of conjugate complex numbers when a quadratic equation with real coefficients has complex roots.

We can check the answer to the example above using the sum and product of the roots:

$$\alpha + \beta = (-2 + i) + (-4 - 2i) = -6 - i = -\frac{b}{a}$$

and $\quad \alpha\beta = (-2 + i)(-4 - 2i) = 10 = \dfrac{c}{a}$

✅ *Exam tip*

Note that it is very easy to make arithmetic mistakes when working with complex numbers, so check your working.

Exercise 1.3b

1 (a) Find the complex numbers $u = x + iy$, $x, y \in \mathbb{R}$, where
   $u^2 = -16 + 30i$

   (b) Hence solve for z the quadratic equation
   $z^2 + (1 + i)z + (4 - 7i) = 0$

2 Solve for z the quadratic equation $z^2 - (3 - i)z + (14 - 5i) = 0$

1.4 The Argand diagram

Learning outcomes

- To represent a complex number on an Argand diagram

You need to know

- The difference between a displacement vector and a position vector
- How to represent the sum and difference of vectors geometrically

The Argand diagram

A complex number, $a + ib$, can be represented on a diagram by using the ordered pair (a, b) to represent a point A on a pair of perpendicular axes, as shown on the diagram.

Then the vector $\overrightarrow{OA}$ represents the complex number $a + ib$

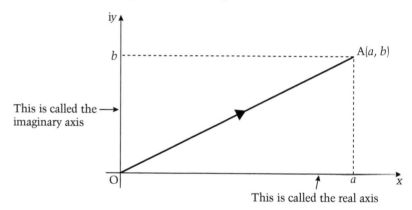

This is called an **Argand diagram**.

Any complex number $x + iy$ can be represented by $\overrightarrow{OP}$ where P is the point (x, y).

The complex number $5 + 3i$ can be represented by $\overrightarrow{OA}$ where A is the point $(5, 3)$.

Any other vector with the same length and direction can also be used to represent $5 + 3i$, for example $\overrightarrow{DE}$ or $\overrightarrow{BC}$.

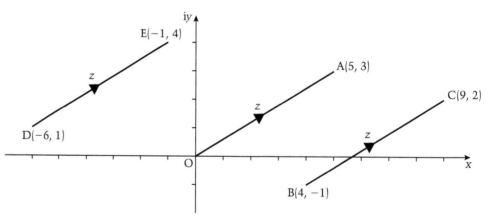

Therefore a complex number can be represented by a displacement vector. It can also be represented by a position vector, when it can also be represented by the point A.

When $z_1 = 5 + 3i$, the vector representing z_1 must be marked with an arrow to show its direction.

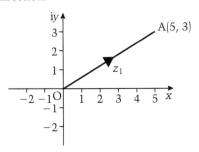

Example

(a) Illustrate on an Argand diagram the points A and B representing the complex numbers $z_1 = 3 - 2i$ and $z_2 = -1 + 2i$, respectively.

(b) On the same diagram illustrate $z_1 + z_2$ and interpret the result in terms of the vectors representing z_1 and z_2.

(a)

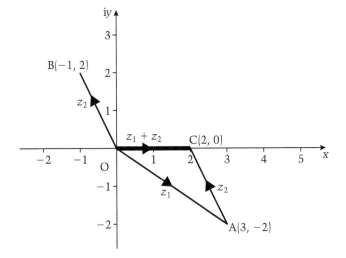

(b) $z_1 + z_2 = (3 - 2i) + (-1 + 2i) = 2$

This is represented by the point C and the vector $\overrightarrow{OC}$.

$\overrightarrow{OC} = \overrightarrow{OA} + \overrightarrow{AC}$ and $\overrightarrow{AC}$ represents z_2.

Therefore the vector representing $z_1 + z_2$ is represented by the sum of the vectors representing z_1 and z_2.

Exercise 1.4

1 Given $z = 3 - 2i$, represent z on an Argand diagram.
On the same diagram represent z^*.

2 Find the square roots of $z = 2i$
Represent z and its two square roots on an Argand diagram.

Learning outcomes

- To define the modulus and argument of a complex number
- To introduce the polar-argument form of a complex number

You need to know

- How to represent a complex number on an Argand diagram

The modulus of a complex number

The point $A(a + ib)$ can be located using the distance, r, of A from the origin O and the angle, θ, that $\overrightarrow{OA}$ makes with the positive x-axis.

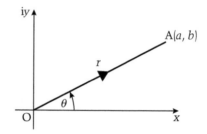

The length of OA, r, is called the modulus of $a + ib$, which is written as $|a + ib|$

$$\therefore \quad |a + ib| = r = \sqrt{a^2 + b^2}$$

For example, the modulus of $3 - 4i$ is $\sqrt{3^2 + 4^2} = 5$

The argument of a complex number

The angle θ is called the argument of $a + ib$ and is written as $\arg(a + ib)$

$$\therefore \quad \arg(a + ib) = \theta \quad \text{where} \quad \tan\theta = \frac{b}{a} \quad \text{and} \quad -\pi < \theta \leqslant \pi$$

To find the argument of a complex number, draw it on an Argand diagram so you can see which quadrant it is in.

For example, the complex numbers $4 + 3i$, $-4 + 3i$, $-4 - 3i$ and $4 - 3i$ are drawn in the diagrams below.

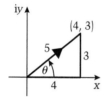 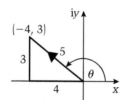 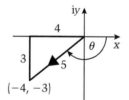 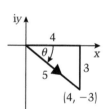

$4 + 3i$ is in the first quadrant, so $\tan\theta = \frac{3}{4} \Rightarrow \theta = 0.644\,\text{rad}$

$-4 + 3i$ is in the second quadrant, so
$$\tan\theta = -\frac{3}{4} \Rightarrow \theta = \pi - \tan^{-1}\frac{3}{4} = 2.50\,\text{rad}$$

$-4 - 3i$ is in the third quadrant, so θ is negative and obtuse.
Therefore $\tan\theta = \frac{3}{4} \Rightarrow \theta = -\pi + \tan^{-1}\frac{3}{4} = -2.50\,\text{rad}$

$4 - 3i$ is in the fourth quadrant, so θ is negative and acute.
Therefore $\tan\theta = -\frac{3}{4} \Rightarrow \theta = -0.644\,\text{rad}$

The polar-argument form of a complex number

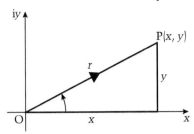

In the diagram above, $x + iy$ is any complex number and we can see that

$$x = r\cos\theta \text{ and } y = r\sin\theta$$

Therefore $x + iy$ can be written as $r\cos\theta + ir\sin\theta$

Hence $x + iy \equiv r(\cos\theta + i\sin\theta)$

$r(\cos\theta + i\sin\theta)$ is called the **polar-argument** form of a complex number. Note that the '+' sign is important: $3(\cos\frac{\pi}{3} - i\sin\frac{\pi}{3})$ is *not* in polar-argument form but can be converted as $-\sin\frac{\pi}{3} = \sin\left(-\frac{\pi}{3}\right)$ and $\cos\frac{\pi}{3} = \cos\left(-\frac{\pi}{3}\right)$

$$\therefore \quad 3\left(\cos\frac{\pi}{3} - i\sin\frac{\pi}{3}\right) = 3\left(\cos\left(-\frac{\pi}{3}\right) + i\sin\left(-\frac{\pi}{3}\right)\right)$$

It is easy to convert between the two forms, as the following worked examples show.

Example

Express $-1 + i$ in the form $r(\cos\theta + i\sin\theta)$

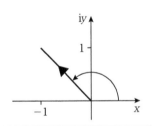

$-1 + i \Rightarrow r = \sqrt{1 + 1} = \sqrt{2}$, and from the diagram, $\theta = \dfrac{3\pi}{4}$

Therefore $-1 + i \equiv \sqrt{2}\left(\cos\dfrac{3\pi}{4} + i\sin\dfrac{3\pi}{4}\right)$

Example

Express $3\left(\cos\left(-\dfrac{\pi}{6}\right) + i\sin\left(-\dfrac{\pi}{6}\right)\right)$ in the form $x + iy$

$\cos\left(-\dfrac{\pi}{6}\right) = \dfrac{\sqrt{3}}{2}$ and $\sin\left(-\dfrac{\pi}{6}\right) = -\dfrac{1}{2}$

$\therefore \quad 3\left(\cos\left(-\dfrac{\pi}{6}\right) + i\sin\left(-\dfrac{\pi}{6}\right)\right) = \dfrac{3\sqrt{3}}{2} - \dfrac{3i}{2}$

Exercise 1.5

1 Find the modulus and argument of each of the following complex numbers.

   (a) $1 - i$ (b) -4

   (c) $2i$ (d) $-3 + 4i$

   (e) $i(1 - i)$

2 Express each of the following complex numbers in the form $x + iy$

   (a) $2\left(\cos\dfrac{\pi}{3} + i\sin\dfrac{\pi}{3}\right)$ (b) $\cos\pi + i\sin\pi$

   (c) $3\left(\cos\dfrac{2\pi}{3} + i\sin\dfrac{2\pi}{3}\right)$

Graphical representation of operations on complex numbers

The sum of two complex numbers

The complex numbers z_1 and z_2 are represented on the Argand diagram by the vectors $\overrightarrow{OA}$ and $\overrightarrow{OB}$ respectively.

Using vector addition we can see that $z_1 + z_2$ is represented by $\overrightarrow{OC}$, where OC is a diagonal of the parallelogram OACB.

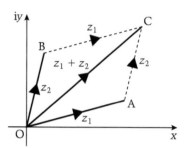

The difference of two complex numbers

Using the same notation as above and using vector subtraction, we can see that $z_1 - z_2$ is represented by $\overrightarrow{BA}$, where BA is the other diagonal of the parallelogram OACB.

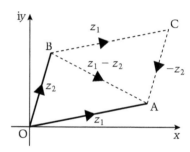

The product of two complex numbers

When $z_1 = r_1(\cos\theta_1 + i\sin\theta_1)$ and $z_2 = r_2(\cos\theta_2 + i\sin\theta_2)$,

then $z_1 z_2 = r_1 r_2(\cos\theta_1 + i\sin\theta_1)(\cos\theta_2 + i\sin\theta_2)$

$$= r_1 r_2(\cos\theta_1\cos\theta_2 - \sin\theta_1\sin\theta_2 + i(\sin\theta_1\cos\theta_2 + \cos\theta_1\sin\theta_2))$$

$$= r_1 r_2(\cos(\theta_1 + \theta_2) + i\sin(\theta_1 + \theta_2))$$

Therefore $|z_1 z_2| = |z_1| \times |z_2|$ and $\arg(z_1 z_2) = \arg z_1 + \arg z_2$

Therefore when $z_1 z_2$ is represented on an Argand diagram, we can see that when z_1 is multiplied by z_2, it is enlarged by a scale factor $|z_2|$ and rotated by an angle θ_2.

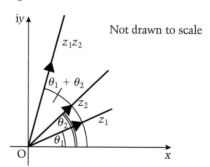

The quotient of two complex numbers

When $z_1 = r_1(\cos\theta_1 + i\sin\theta_1)$ and $z_2 = r_2(\cos\theta_2 + i\sin\theta_2)$, then

$$\frac{z_1}{z_2} = \frac{r_1(\cos\theta_1 + i\sin\theta_1)}{r_2(\cos\theta_2 + i\sin\theta_2)}$$

$$= \frac{r_1}{r_2} \times \frac{(\cos\theta_1 + i\sin\theta_1)(\cos\theta_2 - i\sin\theta_2)}{(\cos\theta_2 + i\sin\theta_2)(\cos\theta_2 - i\sin\theta_2)}$$

$$= \frac{r_1}{r_2} \times \frac{(\cos\theta_1\cos\theta_2 + \sin\theta_1\sin\theta_2) + i(\sin\theta_1\cos\theta_2 - \cos\theta_1\sin\theta_2)}{\cos^2\theta_2 + \sin^2\theta_2}$$

$$= \frac{r_1}{r_2} \times \frac{\cos(\theta_1 - \theta_2) + i\sin(\theta_1 - \theta_2)}{1}$$

$$= \frac{r_1}{r_2}(\cos(\theta_1 - \theta_2) + i\sin(\theta_1 - \theta_2))$$

Therefore $\left|\dfrac{z_1}{z_2}\right| = \dfrac{|z_1|}{|z_2|}$ **and** $\arg\left(\dfrac{z_1}{z_2}\right) = \arg z_1 - \arg z_2$

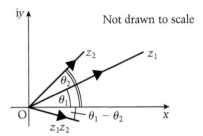

Therefore when $\dfrac{z_1}{z_2}$ is represented on an Argand diagram, we can see that when z_1 is divided by z_2, it is enlarged by a scale factor $\dfrac{1}{|z_2|}$ and rotated by an angle $-\theta_2$.

Example
Find the modulus and argument of $(1 + i)(1 - i\sqrt{3})$

$|1 + i| = \sqrt{2}$ and $\arg(1 + i) = \dfrac{\pi}{4}$;

$|1 - i\sqrt{3}| = 2$ and $\arg(1 - i\sqrt{3}) = -\dfrac{\pi}{3}$

$\therefore \ |(1 + i)(1 - i\sqrt{3})| = |1 + i| \, |1 - i\sqrt{3}| = \sqrt{2} \times 2 = 2\sqrt{2}$

and $\arg(1 + i)(1 - i\sqrt{3}) = \arg(1 + i) + \arg(1 - i\sqrt{3}) = \dfrac{\pi}{4} - \dfrac{\pi}{3} = -\dfrac{\pi}{12}$

Exercise 1.6

1 Given that $z_1 = -2 + 2i\sqrt{3}$ and $z_2 = 2 - 2i$, find the modulus and argument of $\dfrac{z_1}{z_2}$ and hence illustrate z_1, z_2 and $\dfrac{z_1}{z_2}$ on an Argand diagram.

2 Using $z = r(\cos\theta + i\sin\theta)$, prove that $z^2 = r^2(\cos 2\theta + i\sin 2\theta)$
   Hence find the two square roots of $\sqrt{3} - i$

3 Given $z = 1 + i$,
   (a) Express z in polar-argument form.
   (b) Represent z, iz and $\dfrac{1}{iz}$ on an Argand diagram.
   (c) The points A, B and C represent z, iz and $\dfrac{1}{iz}$ respectively.
   Find the complex number represented by $\overrightarrow{BC}$ in the form $a + ib$

1.7   De Moivre's theorem

Learning outcomes

- To state and prove De Moivre's theorem
- To use De Moivre's theorem
- To introduce the exponential form of a complex number

You need to know

- The method of proof by induction
- The meaning of z
- The properties of $\sin\theta$ and $\cos\theta$
- The compound angle and Pythagorean trig identities

Did you know?

Abraham De Moivre (1667–1754) was born in France, but moved to England because of religious intolerance. He was one of the many mathematicians who contributed to the huge advances made in the study of mathematics of that time. He is remembered now mainly because of his theorem. However, he also contributed a great deal to the study of analytic geometry and probability.

De Moivre's theorem

De Moivre's theorem states that

$$(\cos\theta + i\sin\theta)^n = \cos n\theta + i\sin n\theta \text{ for all integral values of } n.$$

De Moivre's theorem is important because it links complex numbers and trigonometry.

Proof by induction of De Moivre's theorem

(To remind yourself of this method of proof, go to Unit 1, Topic 1.6.)

$$(\cos\theta + i\sin\theta)^2 = \cos^2\theta - \sin^2\theta + 2i\sin\theta\cos\theta$$
$$= \cos 2\theta + i\sin 2\theta$$

$\therefore \quad (\cos\theta + i\sin\theta)^n = \cos n\theta + i\sin n\theta \qquad$ when $n = 2$

Assume that $(\cos\theta + i\sin\theta)^n = \cos n\theta + i\sin n\theta$ when $n = k$,

i.e. $\quad (\cos\theta + i\sin\theta)^k = \cos k\theta + i\sin k\theta$

then $\quad (\cos\theta + i\sin\theta)^{k+1}$

$$= (\cos k\theta + i\sin k\theta)(\cos\theta + i\sin\theta)$$
$$= \cos k\theta\cos\theta - \sin k\theta\sin\theta + i(\sin k\theta\cos\theta + \cos k\theta\sin\theta)$$
$$= \cos(k+1)\theta + i\sin(k+1)\theta$$

Therefore if De Moivre's theorem is true when $n = k$ it is also true when $n = k + 1$

We have shown that De Moivre's theorem is true when $n = 2$ so it is also true when $n = 3$

It follows that De Moivre's theorem is true for all positive integer values of n.

Now consider $(\cos\theta + i\sin\theta)^{-n}$ where n is a positive integer.

$$(\cos\theta + i\sin\theta)^{-n}$$

$$= \{(\cos\theta + i\sin\theta)^n\}^{-1}$$

$$= (\cos n\theta + i\sin n\theta)^{-1} \qquad\qquad \textbf{Using the result above}$$

$$= \frac{1}{\cos n\theta + i\sin n\theta}$$

$$= \frac{(\cos n\theta - i\sin n\theta)}{(\cos n\theta + i\sin n\theta)(\cos n\theta - i\sin n\theta)}$$

$$= \frac{\cos(-n\theta) + i\sin(-n\theta)}{\cos^2 n\theta + \sin^2 n\theta} \qquad \cos n\theta = \cos(-n\theta) \text{ and } -\sin n\theta = \sin(-n\theta)$$

$$= \cos(-n\theta) + i\sin(-n\theta) \qquad\qquad \cos^2 A + \sin^2 A = 1$$

i.e. $(\cos\theta + i\sin\theta)^{-n} = \cos(-n\theta) + i\sin(-n\theta)$

Therefore De Moivre's theorem is true for all integer values of n, positive and negative.

Example

Use De Moivre's theorem to find $(1 + i\sqrt{3})^5$ in the form $a + ib$

First express $1 + i\sqrt{3}$ in polar-argument form:

$$|1 + i\sqrt{3}| = 2 \text{ and } \arg(1 + i\sqrt{3}) = \frac{\pi}{3}$$

$$\Rightarrow \quad 1 + i\sqrt{3} = 2\left(\cos\frac{\pi}{3} + i\sin\frac{\pi}{3}\right)$$

$$\therefore \quad (1 + i\sqrt{3})^5 = 2^5\left(\cos\frac{\pi}{3} + i\sin\frac{\pi}{3}\right)^5$$

$$= 32\left(\cos\frac{5\pi}{3} + i\sin\frac{5\pi}{3}\right)$$

$$= 32\left(\frac{1}{2} - i\frac{\sqrt{3}}{2}\right)$$

$$= 16 - 16i\sqrt{3}$$

Example

When $z = \frac{1}{\sqrt{2}} + i\frac{1}{\sqrt{2}}$ use De Moivre's theorem to show that $z^3 + \frac{1}{z^3}$ is real and find its value.

First express z in polar-argument form:

$$z = \frac{1}{\sqrt{2}} + i\frac{1}{\sqrt{2}}$$

$$= \left(\cos\frac{\pi}{4} + i\sin\frac{\pi}{4}\right)$$

$$\therefore \quad z^3 = \left(\cos\frac{\pi}{4} + i\sin\frac{\pi}{4}\right)^3$$

$$= \left(\cos\frac{3\pi}{4} + i\sin\frac{3\pi}{4}\right) \qquad \text{De Moivre's theorem}$$

$$\text{and} \quad \frac{1}{z^3} = z^{-3} = \left(\cos\frac{\pi}{4} + i\sin\frac{\pi}{4}\right)^{-3}$$

$$= \cos\left(-\frac{3\pi}{4}\right) + i\sin\left(-\frac{3\pi}{4}\right) \qquad \text{De Moivre's theorem}$$

$$\therefore \quad z^3 + \frac{1}{z^3} = \left(\cos\frac{3\pi}{4} + i\sin\frac{3\pi}{4}\right) + \left(\cos\left(-\frac{3\pi}{4}\right) + i\sin\left(-\frac{3\pi}{4}\right)\right)$$

$$= \left(\cos\frac{3\pi}{4} + i\sin\frac{3\pi}{4}\right) + \left(\cos\frac{3\pi}{4} - i\sin\frac{3\pi}{4}\right)$$

$$= 2\cos\frac{3\pi}{4} \text{ which is real}$$

$$= 2 \times \left(-\frac{1}{\sqrt{2}}\right)$$

$$= -\sqrt{2}$$

Example

Given that $z = \cos\theta + i\sin\theta$, show that

$z + \dfrac{1}{z} = 2\cos\theta$ and $z - \dfrac{1}{z} = 2i\sin\theta$

From De Moivre's theorem,

$z^{-1} = \cos(-\theta) + i\sin(-\theta)$

$\qquad = \cos\theta - i\sin\theta$

$\therefore \qquad z + \dfrac{1}{z} = (\cos\theta + i\sin\theta) + (\cos\theta - i\sin\theta)$

$\qquad\qquad = 2\cos\theta$

and $\quad z - \dfrac{1}{z} = (\cos\theta + i\sin\theta) - (\cos\theta - i\sin\theta)$

$\qquad\qquad = 2i\sin\theta$

Notice that when $z = \cos\theta + i\sin\theta$,

$\dfrac{1}{z} = z^{-1} = \cos\theta - i\sin\theta$

i.e. $\dfrac{1}{z} = z^{\star}$

The result from the example above can be extended to give

$$z^n + \frac{1}{z^n} = 2\cos n\theta \quad \text{and} \quad z^n - \frac{1}{z^n} = 2i\sin n\theta$$

These results can be used to prove some trigonometric identities.

Example

Prove that $\cos 3\theta \equiv 4\cos^3\theta - 3\cos\theta$

Starting with $z = \cos\theta + i\sin\theta$

$\qquad \left(z + \dfrac{1}{z}\right)^3 = z^3 + 3z + \dfrac{3}{z} + \dfrac{1}{z^3}$

$\qquad\qquad\qquad = \left(z^3 + \dfrac{1}{z^3}\right) + 3\left(z + \dfrac{1}{z}\right) \qquad [1]$

Using $z = \cos\theta + i\sin\theta$ and the result above gives

$\qquad z + \dfrac{1}{z} = 2\cos\theta$ and $z^3 + \dfrac{1}{z^3} = 2\cos 3\theta$

$[1] \Rightarrow \quad (2\cos\theta)^3 = 2\cos 3\theta + 6\cos\theta$

$\quad \Rightarrow \quad\;\; 4\cos^3\theta = \cos 3\theta + 3\cos\theta$

$\therefore \quad \cos 3\theta \equiv 4\cos^3\theta - 3\cos\theta$

De Moivre's theorem can also be used to simplify expressions.

Example

Simplify $\dfrac{\cos\theta - \mathrm{i}\sin\theta}{\cos 3\theta - \mathrm{i}\sin 3\theta}$

Using $\dfrac{1}{z} = z^\star$

$$\dfrac{1}{\cos 3\theta - \mathrm{i}\sin 3\theta} = \cos 3\theta + \mathrm{i}\sin 3\theta$$

$\therefore \quad \dfrac{\cos\theta - \mathrm{i}\sin\theta}{\cos 3\theta - \mathrm{i}\sin 3\theta}$

$\quad = (\cos\theta - \mathrm{i}\sin\theta)(\cos 3\theta + \mathrm{i}\sin 3\theta)$

$\quad = \cos 3\theta\cos\theta + \sin 3\theta\sin\theta + \mathrm{i}(\sin 3\theta\cos\theta - \cos 3\theta\sin\theta)$

$\quad = \cos 2\theta + \mathrm{i}\sin 2\theta$

The exponential form of a complex number

Euler's formula states that $\cos\theta + \mathrm{i}\sin\theta = \mathrm{e}^{\mathrm{i}\theta}$

Therefore $z = r(\cos\theta + \mathrm{i}\sin\theta)$ can be written as $z = r\mathrm{e}^{\mathrm{i}\theta}$

(Euler's formula is proved in Topic 2.8.)

De Moivre's theorem can sometimes be easier to apply using the exponential form,

for example when $z = 2\mathrm{e}^{\mathrm{i}\frac{\pi}{4}}$,

then, using the laws of indices, $z^3 = 8\left(\mathrm{e}^{\mathrm{i}\frac{\pi}{4}}\right)^3 = 8\mathrm{e}^{\mathrm{i}\frac{3\pi}{4}}$

The following use of Euler's formula gives an interesting equation that links a combination of irrational numbers and an imaginary number to an integer.

$$\mathrm{e}^{\mathrm{i}\pi} = \cos\pi + \mathrm{i}\sin\pi \text{ but } \cos\pi = -1 \text{ and } \sin\pi = 0$$

Therefore $\mathrm{e}^{\mathrm{i}\pi} = -1$

Exercise 1.7

1 Show that $(1 + \mathrm{i})^4$ is real and find its value.

2 Use De Moivre's theorem to prove that $\sin 2\theta = 2\sin\theta\cos\theta$

3 Use De Moivre's theorem to simplify $(\cos 2\theta + \mathrm{i}\sin 2\theta)(\cos\theta + \mathrm{i}\sin\theta)$

4 (a) Express $2\sqrt{2} + 2\mathrm{i}\sqrt{2}$ in the form $r\mathrm{e}^{\mathrm{i}\theta}$
   (b) Use $z^2 = 2\sqrt{2} + 2\mathrm{i}\sqrt{2}$ to find the two square roots of $2\sqrt{2} + 2\mathrm{i}\sqrt{2}$ in the form $r\mathrm{e}^{\mathrm{i}\theta}$

5 Find the value of $(1 + \mathrm{i})^3 + (1 - \mathrm{i})^3$

Learning outcomes

- To investigate the locus of a point in the Argand diagram defined by complex numbers

You need to know

- How to represent sums and differences of complex numbers on an Argand diagram
- The meaning of a line segment and a ray
- The properties of the perpendicular bisector of a line segment
- How to find the points of intersection of curves and lines

Loci

A *locus* (plural *loci*) is a set of points that satisfy a given condition. For example, the locus of points that are at a fixed distance from a fixed point is a circle.

In an Argand diagram, the point P$(x + iy)$ can be anywhere. However, if $z = x + iy$ and we impose the condition $|z| = 4$, then OP is a fixed length of 4 units.

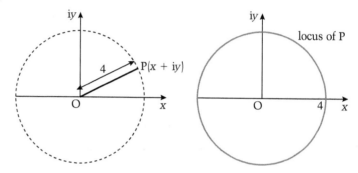

Therefore P is any point on a circle of radius 4 units and centre O.

Any equation of the form $|z| = r$ defines the locus that is a circle of radius r and centre O.

Now consider the equation $|z - z_1| = 4$ where z_1 is the fixed point A$(x_1 + iy_1)$.

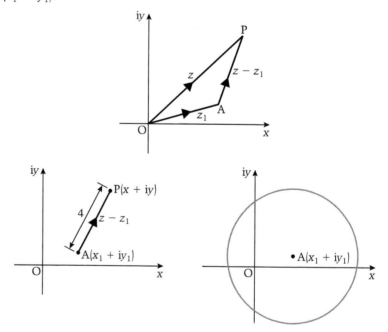

AP $= z - z_1$, so AP is a fixed length of 4 units. Therefore the locus of P is a circle of radius 4 units and centre z_1.

Any equation of the form $|z - z_1| = a$, where a is a real constant, defines a locus that is a circle of radius a and centre z_1.

When you need to work out the locus of a point, it is sensible to start by drawing a diagram.

Example

Sketch on an Argand diagram the locus of points such that $|z - 2i| = 3$

When we compare $|z - 2i| = 3$ with $|z - z_1| = a$

we can see that the locus is a circle whose centre

is the point 2i and whose radius is 3.

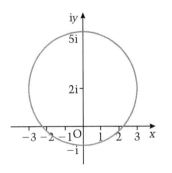

Example

Describe the locus of the points on the Argand diagram given by $|z - z_1| = |z - z_2|$ where z represents the point $P(x + iy)$, $z_1 = 2 + 2i$ and $z_2 = -4 - i$

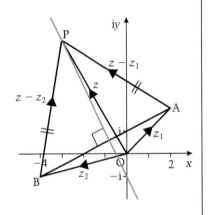

In the diagram, $|z - z_1| = AP$

and $|z - z_2| = BP$

$\therefore$ $|z - z_1| = |z - z_2| \Rightarrow AP = BP$

A point that is equidistant from two fixed points is on the perpendicular bisector of the line segment joining the two points.

Therefore the required locus is the perpendicular bisector of the line segment between $2 + 2i$ and $-4 - i$

This is a particular example of the general result, i.e.

any equation of the form $|z - z_1| = |z - z_2|$ defines a locus that is the perpendicular bisector of the line segment between the points z_1 and z_2.

Example

Describe the locus of points on the Argand diagram given by $\arg(z) = \frac{\pi}{4}$

$\arg(z)$ is the angle that z makes with the positive real axis.

Therefore $\arg(z)$ describes a ray from the origin at an angle of $\frac{\pi}{4}$ to the real axis.

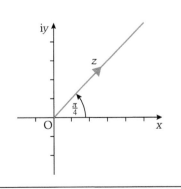

This is another example of a general result, i.e.

> **any equation of the form arg (z) = α describes a ray from the origin at an angle α to the positive direction of the real axis.**

Intersection

To find the points of intersection of two loci, we can convert the equations to Cartesian form. However, this is not always necessary. A diagram will often suggest a simple solution.

Example

Find, in the form $a + ib$, the complex number that satisfies both $|z| = 2$ and $\arg(z) = \frac{\pi}{3}$

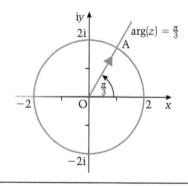

We can see from the diagram that A has a modulus of 2 and an argument of $\frac{\pi}{3}$

Therefore A is the point $2\left(\cos\frac{\pi}{3} + i\sin\frac{\pi}{3}\right)$

i.e. $1 + i\sqrt{3}$

Example

Find the complex numbers that satisfy the equations
$|z - 4| = |z + 2|$ and $|z - 2 - i| = 4$

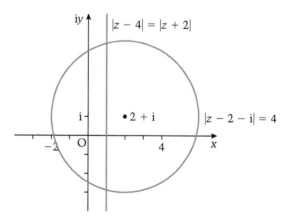

The locus of points that satisfies $|z - 4| = |z + 2|$ is the perpendicular bisector of the line between $x = 4$ and $x = -2$, i.e. $x = 1$

$|z - 2 - i| = 4 \implies |z - (2 + i)| = 4$

and this represents a circle, centre $2 + i$ and radius 4.

To find the complex numbers that satisfy the given equations, we need to find the points on the circle where $x = 1$,

i.e.   where $|1 + iy - 2 - i| = 4$

$\Rightarrow \qquad\qquad |-1 + i(y - 1)| = 4$

$\Rightarrow \qquad\qquad\quad 1 + (y - 1)^2 = 16$

$\Rightarrow \qquad\qquad\quad y^2 - 2y - 14 = 0$

$\Rightarrow \qquad\qquad\qquad\quad y = 1 \pm \sqrt{15}$

Therefore the complex numbers are $1 + i(1 + \sqrt{15})$ and $1 + i(1 - \sqrt{15})$

Region of the Argand diagram

A locus can also be a set of points in a region of the Argand diagram.

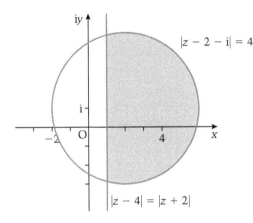

For example, the set of points satisfying $|z - 2 - i| < 4$ lie inside the circle $|z - 2 - i| = 4$

The set of points $|z - 4| < |z + 2|$ lie to the right of the line $x = 1$

The set of points that satisfy both $|z - 2 - i| < 4$ and $|z - 2 - i| = 4$ are contained in the shaded region bounded by the circle and the line.

Exercise 1.8

1   Sketch on an Argand diagram the locus of points for which

(a) $|z| = 2$ (b) $|z - 2i| = 3$

(c) $|z + 2| = 3$ (d) $|z + 2 - 2i| = 2$

2   Sketch on an Argand diagram the locus of points for which

(a) $\arg z = \dfrac{\pi}{4}$ (b) $\arg z = -\dfrac{\pi}{6}$

3   Find the complex numbers satisfied by

(a) $|z| = 5$ and $\arg z = -\dfrac{\pi}{4}$

(b) $|z + 3 - i| = 2$ and $|z| = |z - 2i|$

4   Show on an Argand diagram the set of points for which $x > 3$ and $|z - 2| < 4$

- To differentiate exponential functions

You need to know

- What an exponential function is
- The shape of the curve $y = e^x$
- The meaning of $\dfrac{dy}{dx}$ and $\dfrac{d^2y}{dx^2}$
- How to differentiate multiples, sums, differences, products and quotients of functions
- The chain rule
- The meaning of stationary points

Graphs of $y = a^x$ where $a > 0$

The family of curves with equation $y = a^x$ where $a > 0$, are exponential curves. They are investigated in Unit 1 Topic 1.20.

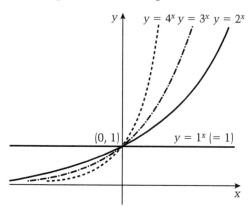

This diagram shows some members of the family. Each of these curves has a property that can be found by drawing accurate plots and by drawing the tangents at some points on the graph, then calculating the gradients of these tangents:

the value of $\dfrac{dy}{dx} \div y$ is constant.

The table below gives approximate values for this constant for $a = 2, 3$ and 4, and the graph shows these values plotted against the value of a.

a	2	3	4
$\dfrac{dy}{dx} \div y$	0.7	1.1	1.4

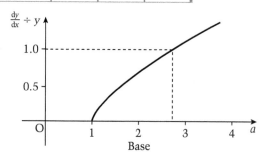

The differential of e^x

The graph shows that there is a number between 2 and 3 for which,

when $y = a^x$, $\dfrac{dy}{dx} \div y = 1$,   i.e. $y = \dfrac{dy}{dx}$

This number is e.

$$\textbf{Therefore when } y = e^x, \; \dfrac{dy}{dx} = e^x$$
$$\textbf{and when } f(x) = e^x, \; f'(x) = e^x$$

The function $f(x) = e^x$ is the only function that is unchanged when it is differentiated.

The differential of $e^{f(x)}$

$e^{f(x)}$ is a composite function so we use the chain rule:

When $y = e^{f(x)}$, $\dfrac{dy}{dx} = \dfrac{dy}{du} \times \dfrac{du}{dx}$ where $u = f(x) \Rightarrow y = e^u$

$\Rightarrow \dfrac{dy}{dx} = e^u \times f'(x) = f'(x)e^{f(x)}$

i.e. the differential of $e^{f(x)}$ is $f'(x)e^{f(x)}$

Example

Find the derivative of

(a) $3e^x$ **(b)** e^{2x+1} **(c)** $x^2 e^{(3x-2)}$ **(d)** $\dfrac{e^x}{\sin x}$

(a) $\dfrac{d}{dx} 3e^x = 3\dfrac{d}{dx} e^x = 3e^x$

(b) $\dfrac{d}{dx} e^{2x+1} = 2e^{2x+1}$ Using the result above

(c) $x^2 e^{(3x-2)}$ is a product so we use $\dfrac{dy}{dx} = u\dfrac{dv}{dx} + v\dfrac{du}{dx}$ where $u = x^2$

    and $v = e^{(3x-2)}$

        $y = x^2 e^{(3x-2)}$

    $\Rightarrow \dfrac{dy}{dx} = (x^2)(3e^{(3x-2)}) + (e^{(3x-2)})(2x)$

         $= (3x^2 + 2x)e^{(3x-2)}$

(d) $\dfrac{e^x}{\sin x}$ is a quotient so we use $\dfrac{dy}{dx} = \dfrac{v\,du - u\,dv}{v^2}$ with $u = e^x$ and

    $v = \sin x$

        $y = \dfrac{e^x}{\sin x}$

    $\Rightarrow \dfrac{dy}{dx} = \dfrac{(\sin x)(e^x) - (e^x)(\cos x)}{\sin^2 x}$

        $= \dfrac{e^x(\sin x - \cos x)}{\sin^2 x}$

Exercise 1.9

1 Find the derivatives of the following functions.

    (a) $5e^x$ **(b)** $e^{-x}\cos x$

    (c) $e^{\sin x}$ **(d)** $\dfrac{e^{2x}}{x^2 - 4}$

2 Given $y = e^x \sin x$ show that $\dfrac{d^2y}{dx^2} - 2\dfrac{dy}{dx} + 2y = 0$

3 Find the coordinates of the stationary point on the curve $y = e^x - x$ and determine its nature.

Learning outcomes

- To differentiate logarithmic functions
- To find the gradients of tangents and normals to curves whose equations are parametric

You need to know

- The meaning of $\ln x$ and its relationship to e^x
- The differential of e^x
- The laws of logarithms
- How to differentiate multiples, sums, differences, products and quotients of functions
- The chain rule
- How to differentiate parametric equations and the meaning of tangents and normals

The differential of $\ln x$

We know that $y = \ln x \iff x = e^y$

Now $\dfrac{d}{dy} e^y = e^y$, therefore $\dfrac{dx}{dy} = e^y = x$

We also know that $\dfrac{dy}{dx} = 1 \div \dfrac{dx}{dy}$ From Unit 1 Topic 3.9

therefore when $y = \ln x$, $\dfrac{dy}{dx} = \dfrac{1}{x}$

i.e.   **when $f(x) = \ln x$, $f'(x) = \dfrac{1}{x}$**

The differential of $\ln f(x)$

$\ln f(x)$ is a composite function so we use the chain rule:

When $y = \ln f(x)$, $\dfrac{dy}{dx} = \dfrac{dy}{du} \times \dfrac{du}{dx}$ where $u = f(x) \Rightarrow y = \ln u$

$\Rightarrow \dfrac{dy}{dx} = \dfrac{1}{u} \times f'(x) = \dfrac{f'(x)}{f(x)}$

i.e.   **the differential of $\ln f(x)$ is $\dfrac{f'(x)}{f(x)}$**

For example, when $y = \ln(x^2 + 1)$, $\dfrac{dy}{dx} = \dfrac{2x}{x^2 + 1}$

The example below shows how the laws of logarithms can be used to simplify the differentiation of many log functions.

Example

Find $\dfrac{dy}{dx}$ when $y = \ln\left(\dfrac{1}{\sqrt{x}}\right)$

$y = \ln\left(\dfrac{1}{\sqrt{x}}\right) = \ln 1 - \ln x^{\frac{1}{2}} = 0 - \frac{1}{2}\ln x$

Therefore $\dfrac{dy}{dx} = -\frac{1}{2}\left(\dfrac{1}{x}\right) = -\dfrac{1}{2x}$

Exercise 1.10a

Find the differential of each function.

1   $\ln 2x$ 2 $\ln x^3$ 3 $\ln(\sin x)$

4   $\ln\left(\dfrac{x}{x^2 + 1}\right)$ 5 $x \ln \sqrt{x^2 + 1}$

Differentiation of parametric equations

We know (from Unit 1 Topic 3.9) that when $y = f(t)$ and $x = g(t)$, where t is a parameter,

$$\dfrac{dy}{dx} = \dfrac{dy}{dt} \div \dfrac{dx}{dt}$$

Equations of tangents and normals

When the equation of a curve is given parametrically, i.e. $x = f(t)$ and $y = g(t)$, we can use $(f(t), g(t))$ as the coordinates of any point on the curve.

This means that we can find the equation of a tangent or normal to the curve at any point on the curve.

For example, when $x = 3t$ and $y = 1 - \dfrac{1}{t}$, the coordinates of any point on the curve are $\left(3t, \; 1 - \dfrac{1}{t}\right)$ and the gradient at any point on the curve is given by $\dfrac{dy}{dx} = \dfrac{dy}{dt} \div \dfrac{dx}{dt} = \dfrac{1}{t^2} \div 3 = \dfrac{1}{3t^2}$

Therefore the equation of the tangent at any point is given by

$$y - \left(1 - \frac{1}{t}\right) = \frac{1}{3t^2}(x - 3t) \qquad\qquad \text{Using } y - y_1 = m(x - x_1)$$

The equation of the normal at any point can also be found: the gradient of the normal is $-3t^2$ so the equation is given by

$$y - \left(1 - \frac{1}{t}\right) = -3t^2(x - 3t)$$

The equation of the tangent and normal at a particular point can be found by substituting the value of t at that point.

Example

The equations of a curve are $x = \cos\theta$ and $y = \theta - \sin\theta$. Find the equation of the normal to this curve in terms of θ. Hence find the equation of the normal at the point where $\theta = \dfrac{\pi}{2}$

$\dfrac{dy}{dx} = \dfrac{1 - \cos\theta}{-\sin\theta} = \dfrac{\cos\theta - 1}{\sin\theta}$ so the gradient of the normal is $\dfrac{\sin\theta}{1 - \cos\theta}$

The equation of the normal is $y - (\theta - \sin\theta) = \dfrac{\sin\theta}{1 - \cos\theta}(x - \cos\theta)$

When $\theta = \dfrac{\pi}{2}$, the equation becomes

$$y - \left(\frac{\pi}{2} - 1\right) = x \;\Rightarrow\; y = x + \frac{\pi}{2} - 1$$

Exercise 1.10b

1   Find, in terms of t, the equation of the tangent to the curve
   $$x = t, \; y = \frac{1}{t}$$
   Hence find the equation of the tangent at the point on the curve where $t = 2$

2   Find, in terms of θ, the equation of the normal to the curve
   $$x = 2\cos\theta, \; y = 3\sin\theta$$
   Hence find the equation of the normal at the point on the curve where $\theta = \dfrac{\pi}{4}$

Learning outcomes

- To describe implicit functions
- To differentiate implicit functions

You need to know

- The product rule and quotient rule for differentiation
- The chain rule

Implicit functions

The equation of some curves, such as $y^2 + xy + x^2y = 2$, are not easy to express in the form $y = f(x)$

A relationship like this is called an ***implicit function*** because $y = f(x)$ is implied by the equation.

Differentiation of implicit functions

The method we use to differentiate an implicit function is to differentiate term by term with respect to x.

The differential of y with respect to x is $\dfrac{dy}{dx}$

To differentiate y^2 with respect to x, we start with $g(y) = y^2$ where $y = f(x)$

Then $g(y) = [f(x)]^2$

This is a composite function so we use the chain rule with the substitution $u = f(x)$

Then $g(y) = u^2$

$$\Rightarrow \quad \frac{d}{dx}(u^2) = \frac{d}{du}(u^2) \times \frac{du}{dx}$$

$$= 2u \times \frac{du}{dx}$$

But $y = u = f(x)$, so $\dfrac{d}{dx}(y^2) = 2y\dfrac{dy}{dx}$

This is a particular example of the general result:

$$\mathbf{\frac{d}{dx}(g(y)) = \left(\frac{d}{dy}\,g(y)\right) \times \frac{dy}{dx}}$$

i.e. to differentiate a function of y with respect to x, differentiate the function with respect to y and multiply by $\dfrac{dy}{dx}$

For example, $\dfrac{d}{dx}(2y^4) = 8y^3\dfrac{dy}{dx}$ and $\dfrac{d}{dx}(\ln y) = \dfrac{1}{y}\dfrac{dy}{dx}$

We can now differentiate any expression involving x and y, term by term, with respect to x.

To differentiate terms such as x^2y we use the product rule,

so $\dfrac{d}{dx}(x^2y) = x^2 \times \dfrac{dy}{dx} + 2x \times y = x^2\dfrac{dy}{dx} + 2xy$

and to differentiate terms such as $\dfrac{x}{y}$ we use the quotient rule,

so $\dfrac{d}{dx}\left(\dfrac{x}{y}\right) = \dfrac{y - x\dfrac{dy}{dx}}{y^2}$

Example

Find $\dfrac{dy}{dx}$ in terms of x and y when $y^2 + xy + x^2y = 2$

Differentiating term by term gives

$$\frac{d}{dx}(y^2) + \frac{d}{dx}(xy) + \frac{d}{dx}(x^2y) = \frac{d}{dx}(2)$$

$$\Rightarrow \quad 2y\frac{dy}{dx} + x\frac{dy}{dx} + y + x^2\frac{dy}{dx} + 2xy = 0$$

$$\Rightarrow \quad \frac{dy}{dx} = -\frac{2xy + y}{2y + x + x^2}$$

Example

Find the gradient of the tangent to the curve $y + xy - y^2 = 6$ at the point $(4, 2)$ on the curve.

We need the value of $\dfrac{dy}{dx}$ when $x = 4$ and $y = 2$

$$y + xy - y^2 = 6$$

$$\Rightarrow \quad \frac{dy}{dx} + y + x\frac{dy}{dx} - 2y\frac{dy}{dx} = 0$$

$$\therefore \quad \frac{dy}{dx} = \frac{y}{2y - x - 1}$$

When $x = 4$ and $y = 2$, $\dfrac{dy}{dx} = -2$

Therefore gradient at the point $(4, 2)$ is -2.

Exercise 1.11

1 Differentiate each equation with respect to x.

  (a) $2x^2 - y^2 = y$ (b) $xe^y = x + y$

  (c) $x\ln(y^2) = 4$ (d) $\dfrac{y^2}{x+1} = y - 1$

2 Find $\dfrac{dy}{dx}$ in terms of x and y when

  (a) $x^2 + xy - y^2 = 6$ (b) $2\cos x + 3\sin y = 4$

3 (a) Find $\dfrac{dy}{dx}$ in terms of x and y when $\sin x + \cos y = 1$

  (b) At the point $\left(\dfrac{\pi}{4}, \alpha\right)$ the gradient of the curve is 1. Find, in the range $0 \leqslant \alpha \leqslant \pi$, the value of α.

1.12 Inverse trigonometric functions

Learning outcomes

- To define the inverse trigonometric functions

You need to know

- The properties and graphs of the sine, cosine and tangent functions
- The meaning of domain and range
- The definition of an inverse function
- The condition for a function to have an inverse

The inverse sine function

The sine function is normally given as $f(x) = \sin x$ for the domain $x \in \mathbb{R}$.

The graph of $f(x) = \sin x$ for this domain is given below.

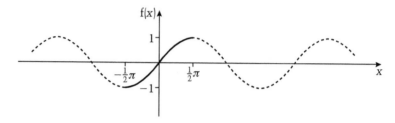

This function does not have an inverse because it is not one-to-one.

However, if we define the function f such that $f(x) = \sin x$ for the domain $-\frac{1}{2}\pi \leqslant x \leqslant \frac{1}{2}\pi$, then the graph of f is the solid line in the graph above. This shows that $f(x)$ is one-to-one and so does have an inverse.

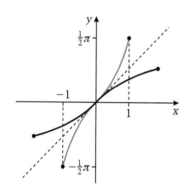

The graph of $y = f^{-1}(x)$ is obtained by reflecting $y = f(x)$ in the line $y = x$ and the equation $y = f^{-1}(x)$ is obtained from $y = f(x)$ by interchanging x and y.

Therefore when $y = \sin x$, $-\frac{1}{2}\pi \leqslant x \leqslant \frac{1}{2}\pi$ the equation of the inverse function is

$\sin y = x$, for $-\frac{1}{2}\pi \leqslant y \leqslant \frac{1}{2}\pi$, i.e. $-1 \leqslant x \leqslant 1$

so y is the angle whose sine is x where $-\frac{1}{2}\pi \leqslant y \leqslant \frac{1}{2}\pi$

The 'angle whose sine is x' is denoted by $\mathbf{\sin^{-1} x}$ (an alternative notation is $\mathbf{\arcsin x}$).

Therefore   when $f(x) = \sin x$, $-\frac{1}{2}\pi \leqslant x \leqslant \frac{1}{2}\pi$

$$f^{-1}(x) = \sin^{-1}x, \ -1 \leqslant x \leqslant 1$$

Note that $\sin^{-1} x$ *is an angle* and that this angle is in the interval $\left[-\frac{1}{2}\pi, \frac{1}{2}\pi\right]$.

The angles in the interval $\left[-\frac{1}{2}\pi, \frac{1}{2}\pi\right]$ are called the ***principal values of*** $\mathbf{\sin^{-1} x}$.

For example, $\sin^{-1}\left(\frac{1}{2}\right)$ is the angle between $-\frac{1}{2}\pi$ and $\frac{1}{2}\pi$ whose sine is $\frac{1}{2}$, so $\sin^{-1}\left(\frac{1}{2}\right) = \dfrac{\pi}{6}$

The inverse cosine function

The function $f(x) = \cos x$, $x \in \mathbb{R}$ is not one-to-one so it does not have an inverse.

However, the function f given by $f(x) = \cos x$, $0 \leqslant x \leqslant \pi$ is one-to-one so f^{-1} exists.

Therefore when $y = \cos x$, $0 \leqslant x \leqslant \pi$ the equation of the inverse function is

$\cos y = x$, for $0 \leqslant y \leqslant \pi$, i.e. $-1 \leqslant x \leqslant 1$

so y is the angle whose cosine is x where $0 \leqslant y \leqslant \pi$

The 'angle whose cosine is x' is denoted by $\cos^{-1} x$ (an alternative notation is **arccos** x).

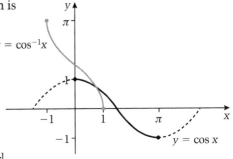

Therefore when $f(x) = \cos x$, $0 \leqslant x \leqslant \pi$

$f^{-1}(x) = \cos^{-1} x$, $-1 \leqslant x \leqslant 1$

Note that $\cos^{-1} x$ *is an angle* and that this angle is in the interval $[0, \pi]$.

The angles in the interval $[0, \pi]$ are called the ***principal values of*** $\cos^{-1} x$.

For example, $\cos^{-1}\left(-\frac{1}{2}\right)$ is the angle between 0 and π whose cosine

is $-\frac{1}{2}$, so $\cos^{-1}\left(-\frac{1}{2}\right) = \dfrac{2\pi}{3}$

The inverse tangent function

When $f(x) = \tan x$, $x \in \mathbb{R}$, f^{-1} does not exist, but when

$f(x) = \tan x$, $-\frac{1}{2}\pi < x < \frac{1}{2}\pi$, f^{-1} does exist.

Therefore when $y = \tan x$, $-\frac{1}{2}\pi < x < \frac{1}{2}\pi$, the equation of the inverse function is

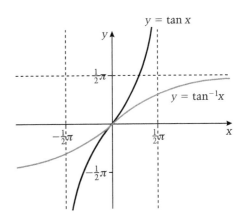

$\tan y = x$ for $-\frac{1}{2}\pi \leqslant y \leqslant \frac{1}{2}\pi$,

so y is the angle whose tangent is x where $x \in \mathbb{R}$

The 'angle whose tan is x' is denoted by $\tan^{-1} x$ (or **arctan** x).

Therefore when $f(x) = \tan x$, $-\frac{1}{2}\pi < x < \frac{1}{2}\pi$

$f^{-1}(x) = \tan^{-1} x$, $x \in \mathbb{R}$

Angles in the interval $\left(-\frac{1}{2}\pi, \frac{1}{2}\pi\right)$ are called the ***principal values of*** $\tan^{-1} x$.

For example, $\tan^{-1}(-1)$ is the angle between $-\frac{1}{2}\pi$ and $\frac{1}{2}\pi$ whose tangent is -1, so $\tan^{-1}(-1) = -\dfrac{\pi}{4}$

Note that the range of $\tan^{-1} x$ is all real values of x whereas the ranges of $\sin^{-1} x$ and $\cos^{-1} x$ are each $[-1, 1]$.

Exercise 1.12

Find the principal value in terms of π of the following.

1 $\cos^{-1}(-1)$

2 $\tan^{-1}(1)$

3 $\sin^{-1}\left(\dfrac{1}{\sqrt{2}}\right)$

4 $\cos^{-1}\left(-\dfrac{\sqrt{3}}{2}\right)$

5 $\tan^{-1}(-\sqrt{3})$

1.13 Differentials of inverse trigonometric functions

The derivative of $\sin^{-1} x$

Let $y = \sin^{-1} x \Rightarrow x = \sin y$

Differentiating $x = \sin y$ with respect to y gives

$$\frac{dx}{dy} = \cos y$$

Therefore using $\dfrac{dy}{dx} = 1 \div \dfrac{dx}{dy}$ gives

$$\frac{dy}{dx} = \frac{1}{\cos y}$$

$$= \frac{1}{\sqrt{1 - \sin^2 y}} \qquad \text{Using } \cos^2 y + \sin^2 y \equiv 1$$

Note that when $y = \sin^{-1} x$, the range of y is $-\frac{1}{2}\pi \leqslant y \leqslant \frac{1}{2}\pi$ and for this range, $\cos y \geqslant 0$

Therefore we only use the positive square root of $1 - \sin^2 y$

But $\sin y = x$, so $\dfrac{dy}{dx} = \dfrac{1}{\sqrt{1 - x^2}}$

$$\text{i.e.} \quad \frac{d}{dx}(\sin^{-1} x) = \frac{1}{\sqrt{1 - x^2}}$$

The derivative of $\cos^{-1} x$

We use the same method to find the derivative of $\cos^{-1} x$

Let $y = \cos^{-1} x \Rightarrow x = \cos y$

Differentiating $x = \cos y$ with respect to y gives

$$\frac{dx}{dy} = -\sin y$$

Therefore using $\dfrac{dy}{dx} = 1 \div \dfrac{dx}{dy}$ gives

$$\frac{dy}{dx} = -\frac{1}{\sin y} \qquad \text{Using } \cos^2 y + \sin^2 y \equiv 1$$

$$= -\frac{1}{\sqrt{1 - \cos^2 y}}$$

When $y = \cos^{-1} x$, the range of y is $0 \leqslant y \leqslant \pi$ and for this range, $\sin y \geqslant 0$

So we only use the positive square root of $1 - \cos^2 y$

But $\cos y = x$, so $\dfrac{dy}{dx} = -\dfrac{1}{\sqrt{1 - x^2}}$

$$\text{i.e.} \quad \frac{d}{dx}(\cos^{-1} x) = -\frac{1}{\sqrt{1 - x^2}}$$

The derivative of $\tan^{-1}x$

The same method again gives the derivative of $\tan^{-1}x$

Let $y = \tan^{-1}x \Rightarrow x = \tan y$

Differentiating $x = \tan y$ with respect to y gives

$$\frac{dx}{dy} = \sec^2 y$$

Therefore using $\frac{dx}{dy} = 1 \div \frac{dx}{dy}$ gives

$$\frac{dx}{dy} = \frac{1}{\sec^2 y} = \frac{1}{1 + \tan^2 y} \qquad \text{Using } 1 + \tan^2 y \equiv \sec^2 y$$

But $\tan y = x$, so $\dfrac{dy}{dx} = \dfrac{1}{1 + x^2}$

$$\text{i.e.} \quad \frac{d}{dx}(\tan^{-1} x) = \frac{1}{1 + x^2}$$

Example

Find the derivative of $\cos^{-1}(3x - 2)$

Let $y = \cos^{-1}(3x - 2)$ and $u = 3x - 2$ so $y = \cos^{-1}u$

Using the chain rule gives $\dfrac{dy}{dx} = -\dfrac{1}{\sqrt{1 - u^2}} \times 3$

$$\Rightarrow \quad \frac{dy}{dx} = -\frac{3}{\sqrt{1 - (3x - 2)^2}}$$

This example is a particular case of the general result, i.e.

$$\frac{d}{dx}(\sin^{-1} f(x)) = \frac{f'(x)}{\sqrt{1 - (f(x))^2}}$$

$$\frac{d}{dx}(\cos^{-1} f(x)) = -\frac{f'(x)}{\sqrt{1 - (f(x))^2}}$$

$$\frac{d}{dx}(\tan^{-1} f(x)) = \frac{f'(x)}{1 + (f(x))^2}$$

Exercise 1.13

Find the derivative of each of the following.

1 $\cos^{-1} 2x$

2 $\tan^{-1}(2x + 1)$

3 $\cos^{-1}(x^2)$

4 $\sin^{-1}(x - 4)$

5 $\tan^{-1}(1 + e^x)$

Learning outcomes

- To find the derivative of a combination of functions

You need to know

- The rules for differentiating products and quotients of functions and composite functions
- How to use logarithms

Summary of differentials

We have found the differentials of a variety of functions in previous topics and in Unit 1.

The results are summarised here.

Standard results

$f(x)$	$f'(x)$
x^n	nx^{n-1}
$\sin x$	$\cos x$
$\cos x$	$-\sin x$
$\tan x$	$\sec^2 x$
e^x	e^x
$\ln x$	$\dfrac{1}{x}$
$\sin^{-1} x$	$\dfrac{1}{\sqrt{1-x^2}}$
$\cos^{-1} x$	$-\dfrac{1}{\sqrt{1-x^2}}$
$\tan^{-1} x$	$\dfrac{1}{1+x^2}$

$gf(x)$	$\dfrac{d}{dx}(gf(x))$
$(ax+b)^n$	$na(ax+b)^{n-1}$
$\sin f(x)$	$f'(x)\cos f(x)$
$\cos f(x)$	$-f'(x)\sin f(x)$
$\tan f(x)$	$f'(x)\sec^2 f(x)$
$e^{f(x)}$	$f'(x)e^{f(x)}$
$\ln f(x)$	$\dfrac{f'(x)}{f(x)}$
$\sin^{-1} f(x)$	$\dfrac{f'(x)}{\sqrt{1-(f(x))^2}}$
$\cos^{-1} f(x)$	$-\dfrac{f'(x)}{\sqrt{1-(f(x))^2}}$
$\tan^{-1} f(x)$	$\dfrac{f'(x)}{1+(f(x))^2}$

Any of these results can be quoted unless you are asked to derive them.

You need to learn these results. When integrating you also need to be able to recognise the function which gives any of these differentials in the table.

For example, given $\dfrac{2x}{x^2+1}$, you need to recognise this as the differential of $\ln(x^2+1)$

General results

$$\frac{d}{dx}f(y) = f'(y)\frac{dy}{dx}$$

$$\frac{dy}{dx} = \frac{dy}{dt} \div \frac{dx}{dt}$$

The use of logarithms

Logarithms were used in Unit 1 to help solve equations with exponents that contained the unknown quantity. The same technique can be used to differentiate functions where the variable is contained in an exponent.

For example, when $y = a^x$, taking logs of both sides gives

$$\ln y = x \ln a$$

Then differentiating with respect to x gives

$$\left(\frac{1}{y}\right)\frac{dy}{dx} = \ln a$$

$$\Rightarrow \quad \frac{dy}{dx} = y \ln a = a^x \ln a$$

Example

Differentiate $x^2 \sin^{-1}(x^2)$

$x^2 \sin^{-1}(x^2)$ is a product so using $y = x^2 \sin^{-1}(x^2)$ with $u = x^2$ and $v = \sin^{-1}(x^2)$,

then $\quad \dfrac{du}{dx} = 2x$ and $\dfrac{dv}{dx} = \dfrac{2x}{\sqrt{1 - x^4}}$

$$\therefore \quad \frac{dy}{dx} = u\frac{dv}{dx} + v\frac{du}{dx} = \frac{2x^3}{\sqrt{1 - x^4}} + 2x \sin^{-1}(x^2)$$

Example

Differentiate $\dfrac{3x}{\ln 5x}$ with respect to x.

$\dfrac{3x}{\ln 5x}$ is a quotient so using $y = \dfrac{3x}{\ln 5x}$ with $u = 3x$ and $v = \ln 5x$

then $\quad \dfrac{du}{dx} = 3$ and $\dfrac{dv}{dx} = \dfrac{1}{x}$

$$\therefore \quad \frac{dy}{dx} = \frac{v\dfrac{du}{dx} - u\dfrac{dv}{dx}}{v^2} = \frac{3\ln 5x - 3}{(\ln 5x)^2}$$

Example

Given $ye^{(2y-1)} = \sin x$ show that $\dfrac{dy}{dx} = \dfrac{y \cot x}{1 + 2y}$

$ye^{(2y-1)}$ is a product, so we use the product rule.

Differentiating with respect to x gives $\quad e^{(2y-1)}\dfrac{dy}{dx} + 2ye^{(2y-1)}\dfrac{dy}{dx} = \cos x$

$$\Rightarrow \quad \frac{dy}{dx} = \frac{\cos x}{(1 + 2y)e^{(2y-1)}} = \frac{y \cos x}{(1 + 2y)\sin x} = \frac{y \cot x}{1 + 2y}$$

Exercise 1.14

1 Find $\dfrac{dy}{dx}$ in terms of x when

   (a) $y = x^{3x}$ **(b)** $y = x\tan^{-1}x$ **(c)** $\sin^{-1}(xy) = x$

2 Find $\dfrac{dy}{dx}$ in terms of x and y when $y = \dfrac{1 - \ln(1-x)^2}{1 + \ln(1+x)^2}$

3 Given $y = x\tan^{-1}x$ show that $x(1 + x^2)\dfrac{dy}{dx} = x^2 + y(1 + x^2)$

Learning outcomes

- To find and use the second differentials of functions

You need to know

- The differentials of standard functions
- The product, quotient and chain rules
- The differential of f(y) with respect to x
- The relationship between e^x and $\ln x$

The second differential of y with respect to x

We met $\dfrac{d^2y}{dx^2}$ in Unit 1. It means the differential of $\dfrac{dy}{dx}$ with respect to x,

i.e. $\dfrac{d}{dx}\left(\dfrac{dy}{dx}\right)$

So, for example, when $\dfrac{dy}{dx} = x^2$,

$\dfrac{d^2y}{dx^2} = $ differential of x^2 with respect to $x = 2x$

> ### Example
>
> If $y = e^{3x}\sin 2x$, find $\dfrac{d^2y}{dx^2}$ in terms of x, simplifying your answer.
>
> $y = e^{3x}\sin 2x \;\Rightarrow\; \dfrac{dy}{dx} = 3e^{3x}\sin 2x + 2e^{3x}\cos 2x$ Using the product rule
>
> $\qquad\qquad\qquad = e^{3x}(3\sin 2x + 2\cos 2x)$
>
> $\qquad \therefore\; \dfrac{d^2y}{dx^2} = 3e^{3x}(3\sin 2x + 2\cos 2x) + e^{3x}(6\cos 2x - 4\sin 2x)$
>
> $\qquad\qquad\quad = e^{3x}(5\sin 2x + 12\cos 2x)$

When an implicit function is differentiated, we often get terms such as $x^2\dfrac{dy}{dx}$

Differentiating such terms will result in a combination of first and second derivatives.

For example, if $y = x^2\dfrac{dy}{dx}$, then using the product rule to differentiate with respect to x gives $\dfrac{dy}{dx} = x^2\dfrac{d^2y}{dx^2} + 2x\dfrac{dy}{dx}$

> ### Example
>
> If $y = \sqrt{3x^2 + 2}$ show that $\left(\dfrac{dy}{dx}\right)^2 + y\dfrac{d^2y}{dx^2} = 3$
>
> $\qquad\qquad y = \sqrt{3x^2 + 2} = (3x^2 + 2)^{\frac{1}{2}}$
>
> $\Rightarrow\quad \dfrac{dy}{dx} = \dfrac{3x}{(3x^2 + 2)^{\frac{1}{2}}}$
>
> Now $y = (3x^2 + 2)^{\frac{1}{2}}$ and we require a relationship that involves y so it is sensible at this stage to substitute y for $(3x^2 + 2)^{\frac{1}{2}}$
>
> This gives $\dfrac{dy}{dx} = \dfrac{3x}{y} \;\Rightarrow\; y\dfrac{dy}{dx} = 3x$
>
> Differentiating with respect to x gives $\left(\dfrac{dy}{dx}\right)\left(\dfrac{dy}{dx}\right) + y\dfrac{d^2y}{dx^2} = 3$
>
> i.e. $\left(\dfrac{dy}{dx}\right)^2 + y\dfrac{d^2y}{dx^2} = 3$

Example

The parametric equations of a curve are $y = t$ and $x = \dfrac{2}{t + 1}$

(a) Show that $2\dfrac{dy}{dx} + x\dfrac{d^2y}{dx^2} = 0$

(b) Hence find $\dfrac{d^2y}{dx^2}$ in terms of t.

(a) We require a relationship that does not involve t, so we start by eliminating t to give a direct relationship between x and y.

$$y = t \text{ and } x = \frac{2}{t + 1} \Rightarrow x = \frac{2}{y + 1} \Rightarrow x(y + 1) = 2$$

$$\therefore \quad x\frac{dy}{dx} + (y + 1) = 0$$

$$\Rightarrow \quad \frac{dy}{dx} + x\frac{d^2y}{dx^2} + \frac{dy}{dx} = 0$$

$$\Rightarrow \quad 2\frac{dy}{dx} + x\frac{d^2y}{dx^2} = 0$$

(b) We can use $x\dfrac{dy}{dx} + (y + 1) = 0$ to give $\dfrac{dy}{dx}$ in terms of t

i.e. $\dfrac{dy}{dx} = -\dfrac{y + 1}{x} = -\dfrac{(t + 1)^2}{2}$

Then $2\dfrac{dy}{dx} + x\dfrac{d^2y}{dx^2} = 0 \Rightarrow -(t + 1)^2 + \left(\dfrac{2}{t + 1}\right)\dfrac{d^2y}{dx^2} = 0$

$$\Rightarrow \frac{d^2y}{dx^2} = \frac{(t + 1)^3}{2}$$

Alternatively we can use $\dfrac{d^2y}{dx^2} = \dfrac{d}{dt}\left(\dfrac{dy}{dx}\right) \times \dfrac{dt}{dx}$ giving

$$\frac{d^2y}{dx^2} = \frac{d}{dt}\left(-\frac{(t + 1)^2}{2}\right) \div \frac{dx}{dt}$$

$$= (-(t + 1)) \div \frac{-2}{(t + 1)^2}$$

$$= \frac{(t + 1)^3}{2}$$

Exercise 1.15

1   If $\tan y = x$, find the value of $\dfrac{d^2y}{dx^2}$ when $y = \dfrac{\pi}{4}$

(Hint: change $\tan y = x$ to $y = \tan^{-1} x$)

2   If $e^y = \sin x$, show that $\dfrac{d^2y}{dx^2} + \left(\dfrac{dy}{dx}\right)^2 + 1 = 0$

3   The parametric equations of a curve are $x = \sin\theta$ and $y = \cos\theta$

(a) Show that $y\dfrac{d^2y}{dx^2} + \left(\dfrac{dy}{dx}\right)^2 + 1 = 0$

(b) Hence find $\dfrac{d^2y}{dx^2}$ in terms of θ

Functions of two or more variables

Many quantities depend on more than one variable.

For example, the profit made by a farmer can depend on the weather, wage costs, the price the farm produce sells for, the cost of transporting the produce to market, and several other variables.

If $z = x^2y$ then z is a function of two variables, x and y, and we write $z = f(x, y)$

If $w = f(x, y, z)$ then w is a function of three variables, x, y, z.

Partial differentiation

The farmer may want to know how profit changes when one of the variables changes while keeping all the others constant, such as when wage costs change.

This is where partial differentiation is useful: if f is a function of x and y, then the partial differential of f with respect to x is found by treating y as a constant.

The **partial differential** of f with respect to x is written as $\frac{\partial f}{\partial x}$

(The formal definition of $\frac{\partial f(xy)}{\partial x}$ is

$$\lim_{h \to 0} \frac{f(x + h, y) - f(x, y)}{h} \text{ where } h = \delta x)$$

For example, if $f(x, y) = xy + 2y^2$

then $\quad \frac{\partial f}{\partial x} = y \qquad\qquad$ This is equivalent to finding $\frac{d}{dx}(ax + b)$

and $\quad \frac{\partial f}{\partial y} = x + 4y \qquad$ This is equivalent to finding $\frac{d}{dy}(ay + 2y^2)$

If you get confused when finding partial derivatives, replace the variables that are treated as constant by letters that look like constants.

Example

If $f(x, y, z) = x^2y + y^2z + xyz$ find

(a) $\frac{\partial f}{\partial x}$ 　　　　　　　　　(b) $\frac{\partial f}{\partial z}$

(a) $\frac{\partial f}{\partial x} = y\frac{d}{dx}x^2 + 0 + yz\frac{d}{dx}x \qquad$ Treating y and z as constants

　　$= 2xy + yz$

(b) $\frac{\partial f}{\partial z} = 0 + y^2\frac{d}{dz}z + xy\frac{d}{dz}z \qquad$ Treating x and y as constants

　　$= y^2 + xy$

Example

If $f(x, y) = xye^x$ find $\dfrac{\partial f}{\partial x}$

Treating y as a constant, we have to differentiate xe^x, so we use the product rule.

$$f(x, y) = xye^x \Rightarrow \dfrac{\partial f}{\partial x} = y\left(\dfrac{d}{dx} xe^x\right) = y(e^x + xe^x)$$

Exercise 1.16a

1 If $f(x, y) = x^2 + y^2$ find $\dfrac{\partial f}{\partial x}$ and $\dfrac{\partial f}{\partial y}$

Hence show that $x\dfrac{\partial f}{\partial x} + y\dfrac{\partial f}{\partial y} = 2f(x, y)$

2 Given $f(x, y, z) = x^2y + y^2z + z^2x$ find $\dfrac{\partial f}{\partial y}$ and $\dfrac{\partial f}{\partial z}$

3 Find $\dfrac{\partial f}{\partial x}$ when

(a) $f(x, y) = x \sin(x + y)$ (b) $f(x, y) = \dfrac{x + y}{x - y}$ (c) $f(x, y) = ye^{(x^2 + 2)}$

Second partial derivative

The symbol $\dfrac{\partial^2 f}{\partial x^2}$ means $\dfrac{\partial}{\partial x}\left(\dfrac{\partial f}{\partial x}\right)$, i.e. find the partial derivative of f with respect to x, then find the partial derivative with respect to x of the result.

Therefore if $f(x, y) = x^3y + xy^3$, then $\dfrac{\partial f}{\partial x} = 3x^2y + y^3$

so $$\dfrac{\partial^2 f}{\partial x^2} = \dfrac{\partial}{\partial x}(3x^2y + y^3) = 6xy$$

With partial differentiation we can also have a mixed second derivative, for example $\dfrac{\partial}{\partial y}\left(\dfrac{\partial f}{\partial x}\right)$ is written as $\dfrac{\partial^2 f}{\partial y \partial x}$

For example when $f(x, y) = x^3y + xy^3$, then

$$\dfrac{\partial f}{\partial x} = 3x^2y + y^3 \qquad\qquad y \text{ constant}$$

So $\dfrac{\partial^2 f}{\partial y \partial x} = \dfrac{\partial}{\partial y}\left(\dfrac{\partial f}{\partial x}\right) = \dfrac{\partial}{\partial y}(3x^2y + y^3) = 3x^2 + 3y^2 \qquad x \text{ constant}$

Exercise 1.16b

1 (a) Given $f(x, y) = e^{(x^2 + y)}$ find

(i) $\dfrac{\partial^2 f}{\partial y^2}$ (ii) $\dfrac{\partial^2 f}{\partial x^2}$ (iii) $\dfrac{\partial^2 f}{\partial y \partial x}$ (iv) $\dfrac{\partial^2 f}{\partial x \partial y}$

(b) State whether the following are true

(i) $\dfrac{\partial^2 f}{\partial x^2} = \dfrac{\partial^2 f}{\partial y^2}$ (ii) $\dfrac{\partial^2 f}{\partial y \partial x} = \dfrac{\partial^2 f}{\partial x \partial y}$

2 Repeat question 1 when $f(x, y) = x \sin(x + y)$

Integration of exponential and logarithmic functions

Learning outcomes

- To integrate exponential functions and logarithmic functions

You need to know

- How to differentiate exponential functions and logarithmic functions
- That integration is the reverse of differentiation
- How to find a definite integral and its interpretation as an area
- The laws of logarithms
- The modulus function

Integration as the reverse of differentiation

When a function is recognised as the differential of a function it can be integrated,

$$\text{so}\quad \frac{d}{dx}\,f(x) = f'(x) \;\Leftrightarrow\; \int f'(x)\,dx = f(x) + c$$

Integration of exponential functions

We know that $\frac{d}{dx}\,e^x = e^x$, therefore $\int e^x\,dx = e^x + c$

We also know that $\frac{d}{dx}\,ae^x = ae^x$ and $\frac{d}{dx}\,e^{(ax+b)} = ae^{(ax+b)}$

Using the chain rule, we also have $\frac{d}{dx}\,e^{f(x)} = f'(x)\,e^{f(x)}$

$$\textbf{therefore}\quad \int ae^x\,dx = ae^x + c \;\text{ and }\; \int e^{(ax+b)}\,dx = \frac{1}{a}\,e^{(ax+b)} + c$$

$$\text{and }\; \int f'(x)e^{f(x)}\,dx = e^{f(x)} + c$$

For example, to find $\int e^{4x}\,dx$ we know that $\frac{d}{dx}\,e^{4x} = 4e^{4x}$

so $\int e^{4x}\,dx = \frac{1}{4}e^{4x} + c$

Example

Evaluate $\displaystyle\int_1^2 4xe^{(x^2-1)}\,dx$

$2x$ is the differential of $x^2 - 1$, so this integral is of the form
$\int f'(x)e^{f(x)}\,dx = e^{f(x)} + c$

$$\therefore\quad \int_1^2 4xe^{(x^2-1)}\,dx = 2\int_1^2 2xe^{(x^2-1)}\,dx = 2\left[e^{(x^2-1)}\right]_1^2 = 2(e^3 - e^0)$$
$$= 2(e^3 - 1)$$

Exercise 1.17a

1 Find

(a) $\displaystyle\int 2e^{2x}\,dx$

(b) $\displaystyle\int e^{(3x-2)}\,dx$

(c) $\displaystyle\int xe^{x^2}\,dx$

(d) $\displaystyle\int (\cos x)e^{\sin x}\,dx$

2 Evaluate (a) $\displaystyle\int_0^1 5e^{4x-1}\,dx$

(b) $\displaystyle\int_0^1 (2x-1)e^{(x^2-x)}\,dx$

Integration of $\dfrac{1}{x}$

If we try to integrate $\frac{1}{x}\,(= x^{-1})$ using $\int x^n\,dx = \frac{1}{n+1}x^{n+1} + c$ we get
$\frac{1}{0}x^0 + c$, which is meaningless.

However, we know that $\frac{d}{dx}\ln x = \frac{1}{x}$. Now $\ln x$ is defined only for $x > 0$, so provided that $x > 0$,

$$\int \frac{1}{x}\,dx = \ln x + c \qquad\qquad [1]$$

When $x < 0$, $\int \frac{1}{x}\,dx = \ln x + c$ is not true, but the function $\frac{1}{x}$ exists for $x < 0$

Also the shaded part of the graph shows that the area represented by

$\int_c^d \frac{1}{x}\,dx$ exists, so it must be possible to integrate $\frac{1}{x}$ for negative values of x.

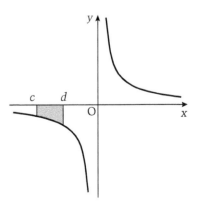

When $x < 0$,

$$-x > 0,\ \int \frac{1}{x}\,dx = \int \frac{-1}{(-x)}\,dx = \ln(-x) + c \qquad [2]$$

[1] and [2] can be combined using $|x|$ to give

$$\int \frac{1}{x}\,dx = \ln|x| + c$$

Integration of $\dfrac{f'(x)}{f(x)}$

We know from Topic 1.9 that $\frac{d}{dx}\ln f(x) = \frac{f'(x)}{f(x)}$, therefore

$$\int \frac{f'(x)}{f(x)}\,dx = \ln|f(x)| + c$$

For example, to find $\int \frac{2}{2x+1}\,dx$, we see that 2 is the differential of $2x + 1$

So $\int \frac{2}{2x+1}\,dx = \ln|2x+1| + c$

Example

Evaluate $\displaystyle\int_2^3 \frac{3x}{x^2-1}\,dx$

$\displaystyle\int_2^3 \frac{3x}{x^2-1}\,dx = \frac{3}{2}\int_2^3 \frac{2x}{x^2-1}\,dx$ and $2x$ is the differential of x^2-1, so

$\displaystyle\int_2^3 \frac{3x}{x^2-1}\,dx = \left[\frac{3}{2}\ln|x^2-1|\right]_2^3 = \frac{3}{2}\ln 8 - \frac{3}{2}\ln 3 = \frac{3}{2}\ln\frac{8}{3}$

Exercise 1.17b

1 Find

(a) $\displaystyle\int \frac{4x}{1-x^2}\,dx$

(b) $\displaystyle\int \frac{e^x}{e^x-1}\,dx$

(c) $\displaystyle\int \frac{\cos x}{\sin x}\,dx$

(d) $\displaystyle\int \frac{1}{x\ln x}\,dx$ $\left(\text{Hint: } \dfrac{1}{x\ln x} = \dfrac{\frac{1}{x}}{\ln x}\right)$

2 Evaluate **(a)** $\displaystyle\int_1^2 \frac{x-2}{x(x-4)}\,dx$ **(b)** $\displaystyle\int_0^1 \frac{e^x}{2e^x-3}\,dx$

Partial fractions

In this section we deal with rational functions, i.e. fractions whose numerators and denominators are polynomials.

We can add or subtract fractions to give a single fraction, for example,

$$\frac{3}{x + 1} + \frac{1}{x - 1} = \frac{3(x - 1) + (x + 1)}{(x + 1)(x - 1)} = \frac{4x - 2}{(x + 1)(x - 1)}$$

The reverse process, i.e. starting with a fraction such as $\dfrac{4x - 2}{(x + 1)(x - 1)}$

and expressing it as the sum or difference of two simpler fractions, is called **decomposing into partial fractions**.

Fractions with linear factors in the denominator

When a fraction is proper (i.e. the highest power of x in the numerator is less than the highest power of x in the denominator) the partial fractions will also be proper.

For example, $\dfrac{2x - 1}{(x + 1)(x - 2)}$ can be expressed as $\dfrac{A}{x + 1} + \dfrac{B}{x - 2}$ where

A and B are numbers.

The worked example shows how the values of A and B can be found.

Example

Express $\dfrac{2x - 1}{(x + 1)(x - 2)}$ in partial fractions.

$$\frac{2x - 1}{(x + 1)(x - 2)} \equiv \frac{A}{x + 1} + \frac{B}{x - 2}$$

First express the right-hand side of this identity as a single fraction over a common denominator.

$$\Rightarrow \quad \frac{2x - 1}{(x + 1)(x - 2)} \equiv \frac{A(x - 2) + B(x + 1)}{(x + 1)(x - 2)}$$

This is an identity: since the denominators are the same then the numerators are also the same.

$$\Rightarrow \quad 2x - 1 \equiv A(x - 2) + B(x + 1)$$

These are two ways of stating the same expression, so we can assign x any value we choose.

Choosing to use $x = 2$ to eliminate A gives

$$3 = 3B \quad \Rightarrow B = 1$$

and using $x = -1$ to eliminate B gives

$$-3 = -3A \Rightarrow A = 1$$

$$\therefore \quad \frac{2x - 1}{(x + 1)(x - 2)} \equiv \frac{1}{x + 1} + \frac{1}{x - 2}$$

Example

Express $\dfrac{x^2 - 2x + 1}{(x - 2)(x + 3)}$ in partial fractions.

$\dfrac{x^2 - 2x + 1}{(x - 2)(x + 3)}$ is improper so we express it as a sum of a polynomial and a proper fraction.

$$\frac{x^2 - 2x + 1}{(x - 2)(x + 3)} \equiv \frac{x^2 - 2x + 1}{x^2 + x - 6} \equiv \frac{(x^2 + x - 6) - 3x + 7}{x^2 + x - 6}$$

$$\equiv 1 + \frac{-3x + 7}{x^2 + x - 6}$$

(This can also be done by dividing $x^2 - 2x + 1$ by $x^2 + x - 6$)

$$\frac{-3x + 7}{x^2 + x - 6} \equiv \frac{A}{x - 2} + \frac{B}{x + 3} \equiv \frac{A(x + 3) + B(x - 2)}{(x - 2)(x + 3)}$$

$\therefore \qquad\qquad -3x + 7 \equiv A(x + 3) + B(x - 2)$

$x = -3 \ \Rightarrow\ 16 = -5B \ \Rightarrow\ B = -\frac{16}{5}$

$x = 2 \quad \Rightarrow \quad 1 = 5A \quad \Rightarrow A = \frac{1}{5}$

$\therefore \quad \dfrac{x^2 - 2x + 1}{(x - 2)(x + 3)} \equiv 1 + \dfrac{1}{5(x - 2)} - \dfrac{16}{5(x + 3)}$

Exercise 1.18a

Express each fraction in partial fractions.

1 $\dfrac{3x}{(x - 1)(x - 2)}$

2 $\dfrac{6}{(x + 1)(2x - 1)}$

3 $\dfrac{2x + 1}{x(x - 1)}$

4 $\dfrac{x^2 - 1}{(x - 2)(x - 3)}$

Fractions with a repeated factor in the denominator

The fraction $\dfrac{x + 1}{(x - 2)^2}$ is a proper fraction and, by adjusting the numerator, can be expressed as the sum of two fractions with numerical numerators.

i.e. $\dfrac{x + 1}{(x - 2)^2} \equiv \dfrac{x - 2 + 3}{(x - 2)^2} \equiv \dfrac{x - 2}{(x - 2)^2} + \dfrac{3}{(x - 2)^2}$

$\equiv \dfrac{1}{(x - 2)} + \dfrac{3}{(x - 2)^2}$

Any fraction whose denominator is a repeated factor can be expressed as two fractions with numerical numerators.

When there are other factors, adjusting the numerator is not easy. The next worked example shows how such a fraction can be decomposed into partial fractions.

Example

Express $\dfrac{2x - 1}{(x - 3)^2(2x + 1)}$ in partial fractions.

$$\frac{2x - 1}{(x - 3)^2(2x + 1)} \equiv \frac{A}{(x - 3)^2} + \frac{B}{(x - 3)} + \frac{C}{(2x + 1)}$$

$$\equiv \frac{A(2x + 1) + B(x - 3)(2x + 1) + C(x - 3)^2}{(x - 3)^2(2x + 1)}$$

$\therefore \qquad 2x - 1 \equiv A(2x + 1) + B(x - 3)(2x + 1) + C(x - 3)^2$

$x = 3 \quad \Rightarrow \quad 5 = 7A \quad$ so $\quad A = \frac{5}{7}$

$x = -\frac{1}{2} \quad \Rightarrow \quad -2 = \frac{49}{4}C \quad$ so $\quad C = -\frac{8}{49}$

The value of B can be found by substituting any value for x (apart from the two values already used).

Choose an easy value: we will use $x = 0$

$x = 0 \quad \Rightarrow \quad -1 = A - 3B + 9C \Rightarrow -1 = \frac{5}{7} - 3B - \frac{72}{49} \quad$ so $\quad B = \frac{4}{49}$

$\therefore \quad \dfrac{2x - 1}{(x - 3)^2(2x + 1)} \equiv \dfrac{5}{7(x - 3)^2} + \dfrac{4}{49(x - 3)} - \dfrac{8}{49(2x + 1)}$

Exercise 1.18b

Express each fraction in partial fractions.

1 $\dfrac{3}{(x - 1)^2(x + 1)}$ $\qquad$ **2** $\dfrac{x^2 + 2}{(x - 2)(x + 1)^2}$ $\qquad$ **3** $\dfrac{x^2 + 1}{x^2(x - 4)}$

Fractions with a quadratic factor in the denominator

Fractions with a quadratic factor in the denominator can also be decomposed into partial fractions.

For example $\dfrac{2x - 1}{(x + 2)(x^2 + 1)}$ is a proper fraction, so its partial fractions will also be proper.

Therefore the partial fraction with denominator $(x^2 + 1)$ will have a linear numerator.

Therefore $\qquad \dfrac{2x - 1}{(x + 2)(x^2 + 1)} \equiv \dfrac{A}{x + 2} + \dfrac{Bx + C}{x^2 + 1}$

$\Rightarrow \quad \dfrac{2x - 1}{(x + 2)(x^2 + 1)} \equiv \dfrac{A(x^2 + 1) + (Bx + C)(x + 2)}{(x + 2)(x^2 + 1)}$

$\Rightarrow \qquad 2x - 1 \equiv A(x^2 + 1) + (Bx + C)(x + 2)$

$x = -2 \quad$ gives $\quad -5 = 5A \qquad\qquad$ so $\quad A = -1$

$x = 0 \quad$ gives $\quad -1 = -1 + 2C \qquad$ so $\quad C = 0$

$x = 1 \quad$ gives $\quad 1 = -2 + (B)(3) \quad$ so $\quad B = 1$

$\therefore \quad \dfrac{2x - 1}{(x + 2)(x^2 + 1)} \equiv -\dfrac{1}{x + 2} + \dfrac{x}{x^2 + 1}$

Repeated quadratic factors

The fraction $\dfrac{x^2 - 3x}{(x^2 + 1)^2}$ has a repeated quadratic factor. By rearranging the numerator this can be expressed as the sum of two fractions,

i.e. $\dfrac{x^2 - 3x}{(x^2 + 1)^2} = \dfrac{(x^2 + 1) - 3x - 1}{(x^2 + 1)^2}$

$= \dfrac{1}{x^2 + 1} - \dfrac{3x + 1}{(x^2 + 1)^2}$

Any repeated quadratic factor can be expressed as the sum of two factors with linear numerators, one with the single quadratic factor and the other with the repeated quadratic factor.

Example

Express $\dfrac{3x + 6}{(x - 1)(x^2 + 2)^2}$ in partial fractions.

$$\dfrac{3x + 6}{(x - 1)(x^2 + 2)^2} \equiv \dfrac{A}{x - 1} + \dfrac{Bx + C}{(x^2 + 2)} + \dfrac{Dx + E}{(x^2 + 2)^2}$$

$$\equiv \dfrac{A(x^2 + 2)^2 + (Bx + C)(x - 1)(x^2 + 2) + (Dx + E)(x - 1)}{(x - 1)(x^2 + 2)^2}$$

$\therefore \quad 3x + 6 \equiv A(x^2 + 2)^2 + (Bx + C)(x - 1)(x^2 + 2) + (Dx + E)(x - 1)$

There are 5 unknowns so we need 5 equations.

$x = 1$ gives $9 = 9A$ so $A = 1$

$x = 0$ gives $6 = 4 - 2C - E$

$\Rightarrow \quad 2C + E = -2$ [1]

$x = -1$ gives $3 = 9 - 6(C - B) - 2(E - D)$

$\Rightarrow \quad -3(C - B) - (E - D) = -3$ [2]

$x = 2$ gives $12 = 36 + 6(C + 2B) + E + 2D$

$\Rightarrow \quad 6(C + 2B) + E + 2D = -24$ [3]

$x = -2$ gives $0 = 36 + 18(2B - C) - 3(E - 2D)$

$\Rightarrow \quad 6(2B - C) - (E - 2D) = -12$ [4]

Solving these equations simultaneously gives $B = -1$, $C = -1$, $D = -3$, $E = 0$

$\therefore \quad \dfrac{3x + 6}{(x - 1)(x^2 + 2)^2} \equiv \dfrac{1}{x - 1} - \dfrac{x + 1}{(x^2 + 2)} - \dfrac{3x}{(x^2 + 2)^2}$

Exercise 1.18c

Express each fraction in partial fractions.

1 $\dfrac{1}{(x + 2)(x^2 + 1)}$ **2** $\dfrac{x^3 + x^2 + 2x}{(x - 1)(x^2 + 1)^2}$ **3** $\dfrac{1}{x^2(x^2 + 1)}$

Learning outcomes

- To use partial fractions to simplify the differentiation of fractions
- To integrate using partial fractions

You need to know

- The differentials of simple functions and of log functions
- How to decompose a rational function into partial fractions
- The integral of $\dfrac{f'(x)}{f(x)}$
- How to find a definite integral
- The laws of logarithms

The use of partial fractions to simplify the differentiation of fractions

We can use the quotient rule to differentiate $\dfrac{1}{(x-1)(x+1)}$ but the simplification of the result is complicated.

When we express $\dfrac{1}{(x-1)(x+1)}$ as partial fractions,

i.e. as $\dfrac{1}{2(x-1)} - \dfrac{1}{2(x+1)}$, then we can differentiate two simpler fractions and the resulting simplification is easier.

$$\therefore \quad \frac{d}{dx}\left(\frac{1}{(x-1)(x+1)}\right) = \frac{d}{dx}\left(\frac{1}{2(x-1)} - \frac{1}{2(x+1)}\right)$$

$$= \frac{d}{dx}\left(\frac{1}{2}(x-1)^{-1}\right) - \frac{d}{dx}\left(\frac{1}{2}(x+1)^{-1}\right)$$

$$= -\frac{1}{2}(x-1)^{-2} + \frac{1}{2}(x+1)^{-2}$$

$$= \frac{1}{2(x+1)^2} - \frac{1}{2(x-1)^2}$$

$$= \frac{(x-1)^2 - (x+1)^2}{2(x-1)^2(x+1)^2}$$

$$= -\frac{4x}{2(x-1)^2(x+1)^2}$$

$$= -\frac{2x}{(x-1)^2(x+1)^2}$$

$$= -\frac{2x}{(x^2-1)^2}$$

Exercise 1.19a

Express each fraction in partial fractions and hence differentiate each fraction.

1 $\dfrac{3x-1}{(x+2)(2x-1)}$

2 $\dfrac{5x}{(x-1)(x-2)(x-3)}$

3 $\dfrac{3x^2-x}{(x+3)(x^2+1)}$

The use of partial fractions in integration

The fraction $\dfrac{1}{(x-1)(x+1)}$ is not recognisable as the differential of a standard function so the integral $\displaystyle\int \frac{1}{(x-1)(x+1)}$ is not obvious, but if we express the fraction in partial fractions, i.e. $\dfrac{1}{(x-1)(x+1)} = \dfrac{1}{2(x-1)} - \dfrac{1}{2(x+1)}$

then $\displaystyle\int \dfrac{1}{(x-1)(x+1)}\,dx = \int \dfrac{1}{2(x-1)}\,dx - \int \dfrac{1}{2(x+1)}\,dx$ and each of

these integrals is recognisable.

$$\therefore \quad \int \dfrac{1}{(x-1)(x+1)}\,dx = \int \dfrac{1}{2(x-1)}\,dx - \int \dfrac{1}{2(x+1)}\,dx$$

$$= \dfrac{1}{2}\ln|x-1| - \dfrac{1}{2}\ln|x+1| + c$$

$$= \dfrac{1}{2}\ln\left|\dfrac{x-1}{x+1}\right| + c$$

Example

Find $\displaystyle\int \dfrac{x^3 + 4x^2 - x}{(x-1)(x+4)}\,dx$

$\dfrac{x^3 + 4x^2 - x}{(x-1)(x+4)}$ is an improper fraction, so first divide the denominator into the numerator:

$$
\begin{array}{r}
x + 1 \\
x^2 + 3x - 4 \overline{\smash{\big)}\, x^3 + 4x^2 - x } \\
\underline{x^3 + 3x^2 - 4x} \\
x^2 + 3x \\
\underline{x^2 + 3x - 4} \\
4
\end{array}
$$

so $\quad \dfrac{x^3 + 4x^2 - x}{(x-1)(x+4)} \equiv x + 1 + \dfrac{4}{(x-1)(x+4)}$

$$\equiv x + 1 + \dfrac{4}{5(x-1)} - \dfrac{4}{5(x+4)}$$

$$\therefore \quad \int \dfrac{x^3 + 4x^2 - x}{(x-1)(x+4)}\,dx = \int (x+1)\,dx + \int \dfrac{4}{5(x-1)}\,dx - \int \dfrac{4}{5(x+4)}\,dx$$

$$= \dfrac{1}{2}x^2 + x + \dfrac{4}{5}\ln|x-1| - \dfrac{4}{5}\ln|x+4| + c$$

$$= \dfrac{x^2}{2} + x + \dfrac{4}{5}\ln\left|\dfrac{x-1}{x+4}\right| + c$$

Exercise 1.19b

1   Use partial fractions to find the following integrals:

(a) $\displaystyle\int \dfrac{3x+4}{x(x+1)}\,dx$

(b) $\displaystyle\int \dfrac{2t}{(t-2)(t+2)}\,dt$

(c) $\displaystyle\int \dfrac{x}{x-2}\,dx$

(d) $\displaystyle\int \dfrac{4x^2 + 3x - 2}{(x+1)(2x+3)}\,dx$

2   Evaluate $\displaystyle\int_{-1}^{0} \dfrac{s^2 + s}{(s-1)(s^2+1)}\,ds$ $\qquad \left(\text{Hint: } \displaystyle\int \dfrac{1}{(1+x^2)} = \tan^{-1}x\right)$

Learning outcomes

- To use substitution to find integrals of some products

You need to know

- That integration is the reverse of differentiation
- The chain rule
- The differentials of the standard functions

Integration using substitution

When $y = gh(x)$ we can use the substitution $u = h(x)$ and the chain rule to find $\frac{d}{dx} gh(x)$ giving

$$\frac{d}{dx} g(u) = g'(u) \frac{du}{dx}$$

$\therefore \quad \int g'(u) \frac{du}{dx} dx = g(u) + c$ \qquad [1]

Now $\quad \int g'(u) du = g(u) + c$ \qquad [2]

Comparing [1] and [2] gives $\qquad \int g'(u) \frac{du}{dx} dx = \int g'(u) du$

When we replace $g(u)$ by $f(u)$ we get $\qquad \int f(u) \frac{du}{dx} dx = \int f(u) du$

Therefore integrating $\left(f(u) \frac{du}{dx} \right)$ with respect to x is equivalent to integrating $f(u)$ with respect to u, i.e.

$$\ldots \frac{du}{dx} dx \equiv \ldots du$$

Note that the relationship above is a pair of equivalent operations. It is *not* an equation nor is it an identity.

For example, to find $\int x^2 (1 - x^3)^{\frac{1}{2}} dx$ using the substitution $u = 1 - x^3$ gives

$$\int x^2 (1 - x^3)^{\frac{1}{2}} dx \quad \Rightarrow \quad \int x^2 u^{\frac{1}{2}} dx$$

Now $\frac{du}{dx} = -3x^2 \quad \therefore \quad \ldots \frac{du}{dx} dx \equiv \ldots du \quad \Rightarrow \quad \ldots (-3x^2) dx \equiv \ldots du$

$$\therefore \quad \int x^2 u^{\frac{1}{2}} dx = \int -\frac{1}{3} u^{\frac{1}{2}} du$$

$$= \left(-\frac{1}{3} \right) \left(\frac{2}{3} \right) u^{\frac{3}{2}} + c = -\frac{2}{9} (1 - x^3)^{\frac{3}{2}} + c$$

This method of substitution is used to integrate a product of functions when one function is the differential of the function 'inside' the other factor. We substitute u for this 'inside' function.

For example, we can use it to find $\int 2 \cos x \sin^3 x \, dx$ because $\cos x$ is the differential of $\sin x$.

We cannot use it to find $\int e^x \cos^2 x \, dx$ because e^x is not the differential of $\cos x$.

Example

Find $\int 2 \cos x \sin^2 x \, dx$

$\cos x$ is the differential of $\sin x$ so we will use the substitution $u = \sin x$

$u = \sin x \quad \Rightarrow \quad \frac{du}{dx} = \cos x, \quad \therefore \quad \ldots \frac{du}{dx} dx \equiv \ldots du \quad \Rightarrow \quad \ldots \cos x \, dx \equiv \ldots du$

$\therefore \quad \int 2 \cos x \sin^2 x \, dx = \int 2u^2 \, du$

$$= \frac{2}{3} u^3 + c = \frac{2}{3} \sin^3 x + c$$

Definite integration using a substitution

When you use the substitution $u = f(x)$ to evaluate a definite integral, you do not need to substitute back to a function of x. You can use $u = f(x)$ to change the limits to corresponding values of u.

After some practice you may find that you can integrate some functions directly without having to make a substitution.

Example

Evaluate $\displaystyle\int_2^3 x\sqrt{x^2 - 1}\, dx$

$2x$ is the differential of $x^2 - 1$, so we will use the substitution $u = x^2 - 1$

$u = x^2 - 1 \quad \Rightarrow \quad \dfrac{du}{dx} = 2x$

$\therefore \quad \ldots \dfrac{du}{dx}\, dx \equiv \ldots du \quad \Rightarrow \quad \ldots 2x\, dx \equiv \ldots du$

$\therefore \quad \displaystyle\int_2^3 x\sqrt{x^2 - 1}\, dx = \frac{1}{2}\int_2^3 2x\sqrt{x^2 - 1}\, dx$

$\qquad\qquad\qquad = \frac{1}{2}\displaystyle\int_{x=2}^{x=3} u^{\frac{1}{2}}\, du$

When $x = 2$, $u = 3$ and when $x = 3$, $u = 8$

$\therefore \quad \displaystyle\int_2^3 x\sqrt{x^2 - 1}\, dx = \frac{1}{2}\int_3^8 u^{\frac{1}{2}}\, du$

$\qquad\qquad\qquad = \frac{1}{2}\left[\frac{2}{3}u^{\frac{3}{2}}\right]_3^8$

$\qquad\qquad\qquad = \frac{1}{3}\left(8^{\frac{3}{2}} - 3^{\frac{3}{2}}\right)$

$\qquad\qquad\qquad = \dfrac{16\sqrt{2} - 3\sqrt{3}}{3}$

Exercise 1.20

1 Use the given substitution to find

    (a) $\displaystyle\int \sin x \sqrt{\cos x}\, dx; \quad u = \cos x$

    (b) $\displaystyle\int \frac{1}{x}(\ln x)^2\, dx; \quad u = \ln x$

2 Use a suitable substitution to find

    (a) $\displaystyle\int e^x \sqrt{e^x - 1}\, dx$

    (b) $\displaystyle\int \cos 2x \sin^2 2x\, dx$

3 Evaluate

    (a) $\displaystyle\int_0^{0.5} x\sqrt{1 - x^2}\, dx$ using the substitution $u = 1 - x^2$

    (b) $\displaystyle\int_0^{\sqrt{3}} \frac{x}{\sqrt{1 + x^2}}\, dx$

Learning outcomes

- To integrate some trig functions

You need to know

- The differentials of the standard trig functions
- That $\frac{d}{dx} f(x) = f'(x)$

  $\Leftrightarrow \int f'(x)\,dx = f(x) + c$
- The laws of logarithms
- The double angle trig identities

Standard trigonometric integrals

From the derivatives of the standard trig functions we know that

$$\int \cos x\,dx = \sin x + c, \quad \int \sin x\,dx = -\cos x + c, \quad \int \sec^2 x\,dx = \tan x + c$$

Using

$$\int \frac{f'(x)}{f(x)}\,dx = \ln|f(x)| + c$$

gives

$$\int \tan x\,dx = \int \frac{\sin x}{\cos x}\,dx = -\ln|\cos x| + c$$

$$= \ln|1| - \ln|\cos x| + c \qquad \ln 1 = 0$$

$$= \ln\left|\frac{1}{\cos x}\right| + c = \ln|\sec x| + c$$

and

$$\int \cot x\,dx = \int \frac{\cos x}{\sin x}\,dx = \ln|\sin x| + c$$

i.e. $\int \tan x\,dx = \ln|\sec x| + c$ and $\int \cot x\,dx = \ln|\sin x| + c$

Using integration by substitution

gives $\int \cos nx\,dx = \frac{1}{n} \sin nx + c, \quad \int \sin nx\,dx = -\frac{1}{n} \cos nx + c,$

and $\int \sin x \cos^n x\,dx = -\frac{1}{n+1} \cos^{n+1} x + c,$

$\int \cos x \sin^n x\,dx = -\frac{1}{n+1} \sin^{n+1} x + c,$

$\int \sec^2 x \tan^n x\,dx = \frac{1}{n+1} \tan^{n+1} x + c$

Even powers of $\sin x$ and $\cos x$

We use the identities for $\cos 2x$ in the forms $\cos^2 x \equiv \frac{1}{2}(1 + \cos 2x)$ and $\sin^2 x \equiv \frac{1}{2}(1 - \cos 2x)$

For example, $\int \sin^2 x\,dx = \int \frac{1}{2}(1 - \cos 2x)\,dx$

$$= \frac{1}{2}x - \frac{1}{4} \sin 2x + c$$

Odd powers of $\sin x$ and $\cos x$

We use the identity $\cos^2 x + \sin^2 x \equiv 1$. For example,

$$\int \sin^3 x\,dx = \int \sin x \,(\sin^2 x)\,dx$$

$$= \int \sin x \,(1 - \cos^2 x)\,dx = \int (\sin x - \sin x \cos^2 x)\,dx$$

$$= -\cos x + \frac{1}{3} \cos^3 x + c$$

Multiple angles

To integrate products such as $\cos 2x \sin 4x$ we can use the *factor formulae* (Topic 2.6 in Unit 1).

For example, to find $\int \cos 2x \sin 4x\, dx$ we can use

$\sin 6x + \sin 2x = 2 \sin 4x \cos 2x$

$\therefore\quad \int \cos 2x \sin 4x\, dx = \frac{1}{2} \int (\sin 6x + \sin 2x)\, dx$

$\qquad\qquad\qquad = -\frac{1}{12} \cos 6x - \frac{1}{4} \cos 2x + c$

A variety of trig functions can be integrated using the ideas given above. The aim is to convert a trigonometric integral to one of the standard forms and/or to reduce the trigonometric function to a number of single trigonometric ratios.

Example

Find $\int \sin^5 \theta\, d\theta$

$\begin{aligned}\sin^5 \theta = \sin \theta \sin^4 \theta &= \sin \theta\, (\sin^2 \theta)^2 \\ &= \sin \theta\, (1 - \cos^2 \theta)^2 \\ &= \sin \theta\, (1 - 2 \cos^2 \theta + \cos^4 \theta)\end{aligned}$

$\therefore\quad \int \sin^5 \theta\, d\theta = \int (\sin \theta - 2 \sin \theta \cos^2 \theta + \sin \theta \cos^4 \theta)\, d\theta$

$\qquad\qquad\qquad = -\cos \theta + \frac{2}{3} \cos^3 \theta - \frac{1}{5} \cos^5 \theta + c$

Example

Evaluate $\int_0^{\frac{\pi}{2}} \sin 2x \sin x\, dx$

$\sin 2x \sin x = -\frac{1}{2} (\cos 3x - \cos x)$

$\therefore\quad \int_0^{\frac{\pi}{2}} \sin 2x \sin x\, dx = \frac{1}{2} \int_0^{\frac{\pi}{2}} (\cos x - \cos 3x)\, dx$

$\qquad\qquad\qquad = \frac{1}{2} \left[\sin x - \frac{1}{3} \sin 3x\right]_0^{\frac{\pi}{2}}$

$\qquad\qquad\qquad = \frac{1}{2} \left(1 - \frac{1}{3}(-1)\right) = \frac{2}{3}$

Exercise 1.21

Find

1 $\int \cos 3x\, dx$

2 $\int \sin \theta \cos^5 \theta\, d\theta$

3 $\int \sec^2 x \tan^3 x\, dx$

4 $\int \cos^3 x \sin^2 x\, dx$

5 $\int \sin x \cos 3x\, dx$

6 $\int \sin 2x \sqrt{1 - 2 \sin^2 x}\, dx$

7 $\int \frac{\cos x}{\sqrt{1 - \sin x}}\, dx$

Evaluate

8 $\int_0^{\frac{\pi}{4}} (\cos 5x \cos 3x)\, dx$

9 $\int_0^{\frac{\pi}{2}} \sin^2 \theta \cos^2 \theta\, d\theta$

1.22 Integration by parts

Learning outcomes

- To integrate a product of functions by parts

You need to know

- The formula for differentiating a product of functions
- The differentials and integrals of standard functions

Integration by parts

We cannot find $\int xe^x\,dx$ using any of the methods introduced so far.

If we start with $\dfrac{d}{dx}uv = v\dfrac{du}{dx} + u\dfrac{dv}{dx}$ and then integrate both sides with respect to x, we get

$$uv = \int v\,\frac{du}{dx}\,dx + \int u\,\frac{dv}{dx}\,dx$$

Rearranging this formula gives

$$\int v\,\frac{du}{dx}\,dx = uv - \int u\,\frac{dv}{dx}\,dx$$

This version of the formula can be used to integrate a product of functions where v and $\dfrac{du}{dx}$ are the two functions; this is called **integrating by parts**.

To use the formula, the right-hand side shows that one of the functions in the product, $\dfrac{du}{dx}$, has to be integrated, and the other function, v, has to be differentiated. When both functions can be integrated, choose as v the function whose differential is the simpler. When only one function can be integrated, then v is the other function. This formula cannot be used when neither function can be integrated.

So, to find $\int xe^x\,dx$, we have two functions that can be integrated. The differential of x is simpler than that of e^x, so we choose $v = x$ and $\dfrac{du}{dx} = e^x$

This gives $u = e^x$ and $\dfrac{dv}{dx} = 1$

Then $\displaystyle\int v\,\frac{du}{dx}\,dx = uv - \int u\,\frac{dv}{dx}\,dx \;\Rightarrow\; \int xe^x\,dx = xe^x - \int (e^x \times 1)\,dx$

$$= xe^x - e^x + c = e^x(x - 1) + c$$

Example

Find $\displaystyle\int x^2 \ln x\,dx$

To use integration by parts to find $\int x^2 \ln x\,dx$ we see that $\ln x$ cannot be integrated but x^2 can. So we choose

$v = \ln x$ and $\dfrac{du}{dx} = x^2 \;\Rightarrow\; \dfrac{dv}{dx} = \dfrac{1}{x}$ and $u = \frac{1}{3}x^3$

$\displaystyle\int v\,\frac{du}{dx}\,dx = uv - \int u\,\frac{dv}{dx}\,dx \;\Rightarrow\; \int x^2 \ln x\,dx = \frac{1}{3}x^3 \ln x - \int \left(\frac{1}{3}x^3\right)\left(\frac{1}{x}\right) dx$

$$= \frac{1}{3}x^3 \ln x - \int \left(\frac{1}{3}x^2\right) dx$$

$$= \frac{1}{3}x^3 \ln x - \frac{1}{9}x^3 + c = \frac{1}{9}x^3(3\ln x - 1) + c$$

Example

Find $\displaystyle\int \ln x\,dx$

We said in the previous example that $\ln x$ cannot be integrated, but using integration by parts with $\ln x = 1 \times \ln x$, we can find $\int \ln x\,dx$

To find $\int (1 \times \ln x)\,dx$, let $v = \ln x$ and $\dfrac{du}{dx} = 1 \;\Rightarrow\; \dfrac{dv}{dx} = \dfrac{1}{x}$ and $u = x$

$\int v\,\dfrac{du}{dx}\,dx = uv - \int u\,\dfrac{dv}{dx}\,dx \;\Rightarrow\; \int \ln x\,dx = x\ln x - \int x\left(\dfrac{1}{x}\right)dx = x\ln x - x + c = x(\ln x - 1) + c$

Example

Find $\int e^x \sin 3x\,dx$

Using $v = e^x$ and $\dfrac{du}{dx} = \sin 3x$ gives $\dfrac{dv}{dx} = e^x$ and $u = -\dfrac{1}{3}\cos 3x$

$$\int e^x \sin 3x\,dx = -\dfrac{1}{3}e^x \cos 3x - \int\left(-\dfrac{1}{3}\cos 3x\right)e^x\,dx$$
$$= -\dfrac{1}{3}e^x \cos 3x + \dfrac{1}{3}\int e^x \cos 3x\,dx$$

Using integration by parts again on $\int e^x \cos 3x\,dx$ with $v = e^x$ and $\dfrac{du}{dx} = \cos 3x$ gives

$$\int e^x \sin 3x\,dx = -\dfrac{1}{3}e^x \cos 3x + \dfrac{1}{3}\left(\dfrac{1}{3}e^x \sin 3x - \dfrac{1}{3}\int e^x \sin 3x\,dx\right)$$
$$= \dfrac{1}{9}e^x \sin 3x - \dfrac{1}{3}e^x \cos 3x - \dfrac{1}{9}\int e^x \sin 3x\,dx$$

The required integral appears on both sides of the equation. Collecting it on the left-hand side gives

$$\dfrac{10}{9}\int e^x \sin 3x\,dx = \dfrac{1}{9}e^x \sin 3x - \dfrac{1}{3}e^x \cos 3x \;\Rightarrow\; \int e^x \sin 3x\,dx = \dfrac{1}{10}e^x(\sin 3x - 3\cos 3x) + c$$

You may find it easier to apply the formula for integration by parts by remembering it in the form

$$\int f(x)\,g(x)\,dx = \left(\int f(x)\right) \times g(x) - \int\left(\left(\int f(x)\right) \times \dfrac{d}{dx}\,g(x)\right)dx$$

Example

Evaluate $\displaystyle\int_0^{\frac{\pi}{2}} x^2 \cos x\,dx$

Using integration by parts, $\int x^2 \cos x\,dx = x^2 \sin x - \int 2x \times \sin x\,dx$

We need to use integration by parts again to find $\int 2x \sin x\,dx$

$\therefore \quad \int x^2 \cos x\,dx = x^2 \sin x - \left(-2x\cos x - \int -2\cos x\,dx\right) = x^2 \sin x + 2x\cos x - 2\sin x + c$

$\therefore \quad \displaystyle\int_0^{\frac{\pi}{2}} x^2 \cos x\,dx = \left[x^2 \sin x + 2x\cos x - 2\sin x\right]_0^{\frac{\pi}{2}} = \left(\dfrac{\pi}{2}\right)^2 - 2$

Exercise 1.22

1 Find

(a) $\displaystyle\int 2xe^{3x}\,dx$

(b) $\displaystyle\int e^x \cos x\,dx$

(c) $\displaystyle\int x\ln x\,dx$

(d) $\displaystyle\int x^2 \sin 2x\,dx$

2 Evaluate

(a) $\displaystyle\int_0^1 \ln(1 + 2x)\,dx$

(b) $\displaystyle\int_0^{\frac{\pi}{4}} 3x \sin 3x\,dx$

(c) $\displaystyle\int_0^{\frac{\pi}{2}} e^{2x} \cos 2x\,dx$

Integration of inverse trigonometric functions

- To find the integrals of the inverse trigonometric functions

You need to know

- The differentials of the inverse trigonometric functions
- How to integrate using substitution
- How to use integration by parts
- Methods for integrating rational functions using partial fractions

The integral of $\sin^{-1}x$

To find $\int \sin^{-1}x\,dx$ we use integration by parts and the same device that we used to find $\int \ln x\,dx$, i.e. we write $\int \sin^{-1}x\,dx$ as $\int (1 \times \sin^{-1}x)\,dx$

Then, using $\int v\dfrac{du}{dx}\,dx = uv - \int u\dfrac{dv}{dx}\,dx$

with $\quad v = \sin^{-1}x \quad$ and $\quad \dfrac{du}{dx} = 1$

gives $\quad \dfrac{dv}{dx} = \dfrac{1}{\sqrt{1-x^2}} \quad$ and $\quad u = x$

$\therefore \quad \int \sin^{-1}x\,dx = x\sin^{-1}x - \int \dfrac{x}{\sqrt{1-x^2}}\,dx$

To find $\int \dfrac{x}{\sqrt{1-x^2}}\,dx$ we can use the substitution

$u = 1 - x^2 \quad \Rightarrow \quad -2x\,dx = du$

$\therefore \quad \int \dfrac{x}{\sqrt{1-x^2}}\,dx = \int \dfrac{-\frac{1}{2}}{u^{\frac{1}{2}}}\,du$

$\qquad\qquad\qquad\qquad = -u^{\frac{1}{2}} + c$

$\qquad\qquad\qquad\qquad = -\sqrt{1-x^2} + c$

(You can also find this integral by sight.)

$\therefore \quad \int \sin^{-1}x\,dx = x\sin^{-1}x - \left(-\sqrt{1-x^2}\right) + c$

i.e. $\quad \int \sin^{-1}x\,dx = x\sin^{-1}x + \sqrt{1-x^2} + c$

The integral of $\cos^{-1}x$

Using a similar method as we used above,

$\int \cos^{-1}x\,dx = \int 1 \times \cos^{-1}x\,dx$

$\qquad\qquad\quad = x\cos^{-1}x - \int \left(x \times \dfrac{-1}{\sqrt{1-x^2}}\right)dx$

$\qquad\qquad\quad = x\cos^{-1}x + \int \dfrac{x}{\sqrt{1-x^2}}\,dx$

$\qquad\qquad\quad = x\cos^{-1}x - \sqrt{1-x^2} + c$

i.e. $\quad \int \cos^{-1}x\,dx = x\cos^{-1}x - \sqrt{1-x^2} + c$

The integral of $\tan^{-1}x$

Using $\int \tan^{-1}x\,dx = \int 1 \times \tan^{-1}x\,dx$ and integration by parts gives

$$\int \tan^{-1}x\,dx = x\tan^{-1}x - \int \frac{x}{1+x^2}\,dx$$

Recognising $\int \frac{x}{1+x^2}\,dx$ as of the form $\frac{1}{2}\int \frac{f'(x)}{f(x)}\,dx = \frac{1}{2}\ln f(x) + c$

gives $\quad \int \frac{x}{1+x^2}\,dx = \frac{1}{2}\ln(1+x^2) + c$

$$\therefore \quad \int \tan^{-1}x\,dx = x\tan^{-1}x - \frac{1}{2}\ln(1+x^2) + c$$

These results can be quoted unless their derivation is asked for, although it is better to remember the method rather than learn the integrals.

Example

Find $\int x\tan^{-1}x\,dx$

Using integration by parts gives

$$\int x\tan^{-1}x\,dx = \frac{1}{2}x^2\tan^{-1}x - \int \frac{x^2}{2(1+x^2)}\,dx$$

Now $\quad \dfrac{x^2}{1+x^2} = \dfrac{1+x^2-1}{1+x^2} = 1 - \dfrac{1}{1+x^2}$

$$\therefore \quad \int \frac{x^2}{2(1+x^2)}\,dx = \frac{1}{2}\int \frac{x^2}{1+x^2}\,dx$$

$$= \frac{1}{2}\left(\int 1\,dx - \int \frac{1}{1+x^2}\,dx\right)$$

$$= \frac{1}{2}x - \frac{1}{2}\tan^{-1}x + c$$

$$\therefore \quad \int x\tan^{-1}x\,dx = \frac{1}{2}x^2\tan^{-1}x - \frac{1}{2}(x - \tan^{-1}x) + c$$

$$= \frac{1}{2}(x^2 + 1)\tan^{-1}x - \frac{1}{2}x + c$$

☑ *Exam tip*

When using integration by parts, it is sensible to check your answer by differentiating it: this should give the function you integrated.

Exercise 1.23

1 Find

(a) $\int \tan^{-1}3x\,dx$ (b) $\int \sin^{-1}2x\,dx$ (c) $\int x^2\tan^{-1}x\,dx$

2 Evaluate

(a) $\int_0^1 \sin^{-1}x\,dx$ (b) $\int_0^{\frac{1}{\sqrt{2}}} \cos^{-1}x\,dx$ (c) $\int_0^{\frac{1}{2}} \tan^{-1}(1+x)\,dx$

Learning outcomes

- To derive and use reduction formulae

You need to know

- How to use integration by parts
- The integrals of standard functions

Finding a reduction formula

If we need to find $\int \sin^4 x \, dx$ we can use the identity $\sin^2 \theta \equiv \frac{1}{2}(1 - \cos 2\theta)$ then square it to give $\sin^4 \theta = \frac{1}{4}(1 - 2\cos 2\theta + \cos^2 2\theta)$.

We can then use the identity $\cos^2 2\theta = \frac{1}{2}(1 + \cos 4\theta)$ to give an integral involving $\cos 2\theta$ and $\cos 4\theta$, but this is tedious if used to find the integral of higher powers of $\sin x$.

To do this we use integration by parts to give a formula that systematically reduces the power to one that we can easily integrate.

For example to find $\int \sin^n x \, dx$, where n is a positive integer, we start by calling this integral I_n

then, by writing $\int \sin^n x \, dx$ as $\int \sin x \, \sin^{n-1} x \, dx$ and using integration by parts, we have

$$I_n = \int \sin^n x \, dx$$

$$= \int \sin x \, \sin^{n-1} x \, dx$$

$$= -\cos x \, \sin^{n-1} x - \int (-\cos x)((n-1)\cos x \sin^{n-2} x) \, dx$$

$$= -\cos x \, \sin^{n-1} x + \int (n-1)(\cos^2 x \sin^{n-2} x) \, dx$$

$$= -\cos x \, \sin^{n-1} x + (n-1) \int ((1 - \sin^2 x) \sin^{n-2} x) \, dx$$

$$= -\cos x \, \sin^{n-1} x + (n-1)\left(\int \sin^{n-2} x \, dx - \int \sin^n x \, dx \right)$$

$$\therefore \quad I_n = -\cos x \, \sin^{n-1} x + (n-1)(I_{n-2} - I_n)$$

$$\Rightarrow \quad nI_n = -\cos x \, \sin^{n-1} x + (n-1)I_{n-2}$$

$$\Rightarrow \quad I_n = -\frac{1}{n}\cos x \, \sin^{n-1} x + \frac{n-1}{n}I_{n-2}$$

This is called a ***reduction formula*** because it reduces an integral involving the power n to one involving the power $n - 2$.

Depending on the function to be integrated, a reduction formula may reduce the power by 1 or by 2.

Using a reduction formula

A reduction formula can be used to systematically reduce the power to one where the integral can be found easily.

To find $\int \sin^6 x \, dx$ we have $I_6 = \int \sin^6 x \, dx$ and using the formula above

$$I_6 = -\frac{1}{6}\cos x \, \sin^5 x + \frac{5}{6}I_4 \qquad \text{[1]}$$

Using the formula again on I_4 gives

$$I_4 = -\frac{1}{4}\cos x \, \sin^3 x + \frac{3}{4}I_2 \qquad \text{[2]}$$

Now $\quad I_2 = \int \sin^2 x \, dx$

$$= \int \tfrac{1}{2} (1 - \cos 2x) \, dx$$

$$= \tfrac{1}{2} x - \tfrac{1}{4} \sin 2x + c$$

From [2], $\quad I_4 = -\tfrac{1}{4} \cos x \sin^3 x + \tfrac{3}{4} \left(\tfrac{1}{2} x - \tfrac{1}{4} \sin 2x + c \right)$

$$= \tfrac{3}{8} x - \tfrac{1}{4} \cos x \sin^3 x - \tfrac{3}{16} \sin 2x + c$$

From [1], $\quad I_6 = -\tfrac{1}{6} \cos x \sin^5 x - \tfrac{5}{24} \cos x \sin^3 x + \tfrac{5}{16} x - \tfrac{5}{16} \cos x \sin x + c$

Using $\sin 2x = 2 \sin x \cos x$

Note that when we found I_2, we introduced a constant of integration, c. As c is an unknown constant, there is no need to multiply it by the fractions in the reduction formulae.

Example

(a) Given $I_n = \int x^n e^x \, dx$ show that $I_n = x^n e^x - n I_{n-1}$

(b) Hence find $\int x^4 e^x \, dx$

(a) $\quad I_n = \int x^n e^x \, dx = x^n e^x - \int n x^{n-1} e^x \, dx$

$$= x^n e^x - n \int x^{n-1} e^x \, dx$$

$$= x^n e^x - n I_{n-1}$$

(b) $I_4 = \int x^4 e^x \, dx$

$\quad$ Now $\quad I_1 = \int x e^x \, dx$

$$= x e^x - \int e^x \, dx$$

$$= e^x (x - 1) + c \qquad \text{Using integration by parts}$$

$\quad$ Using $\quad I_n = x^n e^x - n I_{n-1}$ with $n = 2$ gives

$$I_2 = x^2 e^x - 2 I_1$$

$$= x^2 e^x - 2 e^x (x - 1) + c$$

$$= e^x (x^2 - 2x + 2) + c$$

$\quad$ Similarly $\quad I_3 = x^3 e^x - 3 I_2$

$$= e^x (x^3 - 3x^2 + 6x - 6) + c$$

$\quad$ and $\quad I_4 = x^4 e^x - 4 I_3$

$$= e^x (x^4 - 4x^3 + 12x^2 - 24x + 24) + c$$

We can also find a reduction formula for a definite integral, and in this case the formula is often easier to use.

Example

(a) Given $I_n = \displaystyle\int_0^1 x(x^2 - 1)^n \, dx$ show that $I_n = -\dfrac{n}{n+1} I_{n-1}$

(b) Hence evaluate $\displaystyle\int_0^1 x(x^2 - 1)^5 \, dx$

(a) We need an integral involving $(x^2 - 1)^{n-1}$ so we start by writing I_n as

$$\int_0^1 x(x^2 - 1)(x^2 - 1)^{n-1} \, dx$$

$$\therefore \quad I_n = \int_0^1 x(x^2 - 1)(x^2 - 1)^{n-1} \, dx$$

We can now express this as two integrals, i.e.

$$I_n = \int_0^1 x^3(x^2 - 1)^{n-1} \, dx - \int_0^1 x(x^2 - 1)^{n-1} \, dx$$

$$= \int_0^1 x^3(x^2 - 1)^{n-1} \, dx - I_{n-1}$$

$$\therefore \quad I_n + I_{n-1} = \int_0^1 x^3(x^2 - 1)^{n-1} \, dx$$

If we use integration by parts with $v = (x^2 - 1)^{n-1}$ we will reduce the power to $n - 2$, which we do not want. Therefore we rearrange $x^3(x^2 - 1)^{n-1}$ so that there is a term involving $(x^2 - 1)^{n-1}$ that can be integrated.

Now $\displaystyle\int_0^1 x^3(x^2 - 1)^{n-1} \, dx = \int_0^1 (x^2)\left(x(x^2 - 1)^{n-1}\right) dx$

Using integration by parts with $\dfrac{du}{dx} = x(x^2 - 1)^{n-1}$ gives

$$I_n + I_{n-1} = \left[(x^2)\left(\frac{(x^2 - 1)^n}{2n}\right)\right]_0^1 - \int_0^1 2x\left(\frac{(x^2 - 1)^n}{2n}\right) dx$$

$$= 0 - \frac{1}{n} I_n$$

$$\therefore \quad I_n + I_{n-1} = -\frac{1}{n} I_n$$

$$\Rightarrow I_n = -\frac{n}{n+1} I_{n-1}$$

(b) $\displaystyle\int_0^1 x(x^2 - 1)^5 \, dx = I_5$

Now $I_1 = \displaystyle\int_0^1 x(x^2 - 1) \, dx$

$$= \int_0^1 (x^3 - x) \, dx$$

$$= \left[\frac{x^4}{4} - \frac{x^2}{2}\right]_0^1$$

$$= -\frac{1}{4}$$

Using $I_n = -\dfrac{n}{n+1} I_{n-1}$ gives

$$I_2 = -\tfrac{2}{3} \times I_1$$
$$= \left(-\tfrac{2}{3}\right)\left(-\tfrac{1}{4}\right)$$
$$= \tfrac{1}{6}$$

so $\quad I_3 = -\tfrac{3}{4} I_2$
$$= -\tfrac{3}{4} \times \tfrac{1}{6}$$
$$= -\tfrac{1}{8}$$

and $\quad I_4 = -\tfrac{4}{5} I_3$
$$= -\tfrac{4}{5} \times -\tfrac{1}{8}$$
$$= \tfrac{1}{10}$$

so $\quad I_5 = -\tfrac{5}{6} \times \tfrac{1}{10}$
$$= -\tfrac{1}{12}$$

i.e. $\displaystyle\int_0^1 x(x^2 - 1)^5 \,dx = -\tfrac{1}{12}$

Note that a formula like $I_n = -\dfrac{n}{n+1} I_{n-1}$, where I_1 is known, produces a sequence giving values for $I_2, I_3, I_4, \ldots$

Such a formula is called a ***recurrence relation***.

Exercise 1.24

1 If $I_n = \displaystyle\int \cos^n x \,dx$ show that

$$I_n = \frac{1}{n} \sin x \cos^{n-1} x + \frac{n-1}{n} I_{n-2}, \ n \geqslant 2$$

Hence find $\displaystyle\int \cos^6 x \,dx$

2 Use the reduction formula given in question 1 to show that,

when $I_n = \displaystyle\int_0^{\frac{\pi}{2}} \cos^n x \,dx$,

$$I_n = \frac{n-1}{n} I_{n-2}$$

3 If $I_n = \displaystyle\int_0^1 x^n \sqrt{1-x} \,dx$ show that

$$I_n = \frac{2n}{2n+3} I_{n-1}, \ n > 0$$

Hence find $\displaystyle\int_0^1 x^6 \sqrt{1-x} \,dx$

1.25 The trapezium rule

We have covered a variety of methods to integrate functions. However, there are many indefinite integrals that cannot be found. But there are several methods that can be used to give an approximation for a definite integral when the function involved cannot be integrated. We look at one such method here.

The trapezium rule

The integral $\int_a^b f(x)\,dx$ represents the area between the curve $y = f(x)$, the x-axis and the ordinates $x = a$ and $x = b$.

When a function whose derivative is $f(x)$ cannot be found, we can divide the area into a finite number of vertical strips as shown in the diagram. Joining the tops of the strips as shown gives a set of trapezia.

The sum of the areas of these trapezia then gives an approximate value for $\int_a^b f(x)\,dx$

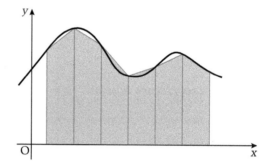

Taking n strips at equal intervals along the x-axis so that each strip is the same width, d, and labelling the vertical sides (i.e. the ordinates) $y_0, y_1, ..., y_n$,

then the area of the first strip is $\frac{1}{2}d\,(y_0 + y_1)$,

the area of the second strip is $\frac{1}{2}d\,(y_1 + y_2)$, and so on.

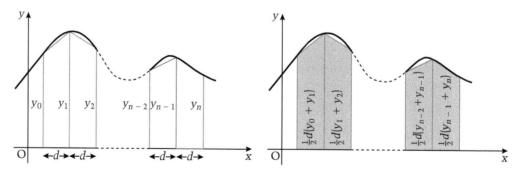

The sum of the areas of all the strips is given by

$$\tfrac{1}{2}d\,(y_0 + y_1) + \tfrac{1}{2}d\,(y_1 + y_2) + ... + \tfrac{1}{2}d\,(y_{n-2} + y_{n-1}) + \tfrac{1}{2}d\,(y_{n-1} + y_n)$$
$$= \tfrac{1}{2}d\,(y_0 + 2y_1 + 2y_2 ... + 2y_{n-2} + 2y_{n-1} + y_n)$$

$$\therefore \quad \int_a^b f(x)\,dx \simeq \tfrac{1}{2}d\,(y_0 + 2y_1 + 2y_2 ... + 2y_{n-2} + 2y_{n-1} + y_n)$$

This formula is called the trapezium rule.
It is easy to remember this formula in words as
half the width of a strip × (first + last + twice all the other) ordinates.

Note that there is one more ordinate than the number of strips.

Note also that the approximation gets better when the width of the strips,
i.e. the value of d, gets smaller.

Example

(a) Find an approximate value for $\int_1^6 x^3 \, dx$ using the trapezium rule with five intervals.

(b) Use a sketch to determine whether your answer is an over-estimate or an under-estimate.

(a) $\int_1^6 x^3 \, dx$ represents the area between $y = x^3$, the x-axis and the ordinates $x = 1$ and $x = 6$

There are five units between $x = 1$ to $x = 6$ so we take our five intervals as one unit wide, i.e. $d = 1$

This gives six ordinates: $y_0 = 1^3 = 1$, $y_2 = 2^3 = 8$, $y_3 = 27$, $y_4 = 64$, $y_5 = 125$, $y_6 = 216$

The trapezium rule gives

$\int_1^6 x^3 \, dx \approx \frac{1}{2}(1 + 216 + 2(8 + 27 + 64 + 125)) = 332.5$

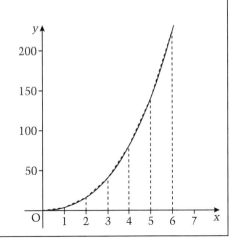

(b) The sketch shows that the area of each trapezium is greater than the area under that part of the curve.

Therefore 332.5 is an over-estimate for the value of $\int_1^6 x^3 \, dx$

Alternatively, $\int_1^6 x^3 \, dx = \left[\frac{1}{4} x^4\right]_1^6 = 323.75$

This is the exact value of the area, confirming that 332.5 is an over-estimate.

Exercise 1.25

1 (a) Use the trapezium rule with five intervals to find an approximate value for $\int_1^2 \frac{1}{x^2} \, dx$

(b) Sketch the graph showing the area represented by $\int_1^2 \frac{1}{x^2} \, dx$

2 (a) Use the trapezium rule with five intervals to find an approximate value for $\int_0^{\frac{1}{2}} (1 - x^2)^{-\frac{1}{2}} \, dx$

(b) Find the exact value of $\int_0^{\frac{1}{2}} (1 - x^2)^{-\frac{1}{2}} \, dx$

(c) Use your answers to (a) and (b) to find an approximate value for π.

Section 1 Practice questions

1 A quadratic equation with real coefficients has one root equal to $3 - 2i$

Write down

(a) the other root of the equation

(b) the equation.

2 (a) Simplify $\dfrac{2 + i}{3 - 2i} - \dfrac{1 - 2i}{3 + 2i}$

(b) If $x + y = i(a - b)$ for real values of x, y, a and b, write down two relationships between x, y, a and b.

3 Find the values of a and b where $z = a + ib$ such that
$$2iz - z^\star(2 - i) = 2z - 2i$$

4 (a) Find the square roots of $11 - 60i$

(b) Hence find the roots of the equation
$$x^2 - (4 + i)x + (1 + 17i) = 0$$

5 You are given that $z = 1 - i$

(a) Express z in the form $r(\cos\theta + i\sin\theta)$

(b) Find the modulus and argument of z^2

(c) Illustrate z and z^2 and $z - z^2$ in an Argand diagram.

6 You are given that $z = \dfrac{\sqrt{15}}{2} + \dfrac{\sqrt{5}}{2}i$

(a) Express z in the form $r(\cos\theta + i\sin\theta)$

(b) Hence find the two square roots of z.

7 Solve the simultaneous equations
$$z + (2 - i)w = 5 + i$$
$$(2 + i)z - 3w = 3 - i$$

8 (a) Using the binomial theorem or otherwise, expand $(\cos\theta + i\sin\theta)^3$

(b) Hence express
   (i) $\cos 3\theta$ in terms of $\cos\theta$
   (ii) $\sin 3\theta$ in terms of $\sin\theta$

9 (a) Describe the locus of points satisfied by
   (i) $|z - 4| = |z + 6|$ (ii) $|z + 1| = 6$

(b) Hence find in the form $a + ib$ the values of z which satisfy the simultaneous equations
$$|z - 4| = |z + 6| \quad \text{and} \quad |z + 1| = 6$$

10 Find the smallest positive value of x for which $y = e^x \sin x$ has a stationary value and determine the nature of that stationary value.

11 Determine the number and nature of stationary points on the curve whose equation is
$$y = \ln\left(\frac{2x}{x^2 + 1}\right)$$

12 The parametric equations of a curve are
$$x = e^t, y = t$$

(a) (i) Find in terms of t the equation of the tangent to the curve at the point (e^t, t).
   (ii) Hence find the equation of the tangent to the curve at the point where $t = 2$

(b) (i) Find the Cartesian equation of the curve.
   (ii) Use the Cartesian equation to find the equation of the tangent to the curve at the point where $x = e^2$

13 Given that $y = e^y \sin x$ show that
$$\frac{dy}{dx} = \frac{e^y \cos x}{1 - y}$$

Hence find a relationship between $\dfrac{d^2y}{dx^2}, \dfrac{dy}{dx}$ and y.

14 The parametric equations of a curve are $x = 12t^2, y = 3t$

Find $\dfrac{dy}{dx}$ and $\dfrac{d^2y}{dx^2}$ in terms of t.

15 Find the derivatives of

(a) $x^{3x}e^x$

(b) $\sin^{-1}(3x - 2)$

(c) $x\tan^{-1}(2x)$

16 Given that $z = x^2 + xy + y^2$, find $\dfrac{\partial z}{\partial x}$ and $\dfrac{\partial z}{\partial y}$

Hence show that $\left(\dfrac{\partial z}{\partial x}\right)^2 + \left(\dfrac{\partial z}{\partial y}\right)^2 = 5z + 3xy$

17 Given that $z = \sin^{-1}\left(\dfrac{x}{y}\right)$,

find $\dfrac{\partial z}{\partial x}, \dfrac{\partial z}{\partial y}$ and $\dfrac{\partial^2 z}{\partial x \partial y}$

18 Evaluate

(a) $\displaystyle\int_0^{\frac{\pi}{4}} e^{\cos 2x} \sin 2x \, dx$

(b) $\displaystyle\int_0^1 \frac{1 - 2e^{2x}}{x - e^{2x}} \, dx$

19 Express $\dfrac{x - 1}{(x - 2)(x^2 + 1)}$ in partial fractions.

Hence find $\dfrac{dy}{dx}$ when $y = \dfrac{x - 1}{(x - 2)(x^2 + 1)}$

20 Express $\dfrac{3x^2 - 6x - 2}{(2x + 1)(x - 1)^2}$ in partial fractions.

Hence evaluate $\displaystyle\int_{-1}^0 \frac{3x^2 - 6x - 2}{(2x + 1)(x - 1)^2} \, dx$

21 Find

(a) $\displaystyle\int \sin^3 x \, dx$

(b) $\displaystyle\int \sin 5x \cos 3x \, dx$

(c) $\displaystyle\int \cos^4 x \, dx$

22 (a) Use the substitution $u = e^x$ to find

$\displaystyle\int \frac{e^x}{e^{2x} + 1} \, dx$

(b) Hence or otherwise find $\displaystyle\int_0^1 \frac{e^x}{e^{2x} + 1} \, dx$

23 Use the substitution $u = 2^x$ to evaluate $\displaystyle\int_1^2 2^x \, dx$

24 Find

(a) $\displaystyle\int x e^x \, dx$

(b) $\displaystyle\int e^x \cos x \, dx$

(c) $\displaystyle\int x \sec^2 x \, dx$

25 Find

(a) $\displaystyle\int e^{2x} \sin x \, dx$

(b) $\displaystyle\int 2x \log_2 x \, dx$

26 Show that

$$\frac{2\cos x}{\cos x + \sin x} \equiv 1 + \frac{\cos x - \sin x}{\cos x + \sin x}$$

Hence evaluate $\displaystyle\int_0^{\frac{\pi}{2}} \frac{2\cos x}{\cos x + \sin x} \, dx$

27 (a) Given $I_n = \displaystyle\int x^n e^{2x} \, dx$ show that

$$I_n = \frac{1}{2} x^n e^{2x} - \frac{n}{2} I_{n-1}$$

(b) Hence find $\displaystyle\int x^3 e^{2x} \, dx$

28 (a) If $I_n = \displaystyle\int_0^{\frac{\pi}{6}} \sec^n x \, dx$ show that

$$I_n = \frac{2^{n-2}}{\sqrt{3}^{\,n-1}(n - 1)} + \frac{n - 2}{n - 1} I_{n-2} \text{ for } n \geqslant 2$$

(b) Hence evaluate $\displaystyle\int_0^{\frac{\pi}{6}} \sec^4 x \, dx$

29 (a) Use the trapezium rule with three intervals to find an approximate value for $\displaystyle\int_0^1 \frac{1}{1 + x^2} \, dx$

(b) Find the exact value of $\displaystyle\int_0^1 \frac{1}{1 + x^2} \, dx$ and use it to determine whether your answer to part (a) is an over-estimate or an under-estimate.

30

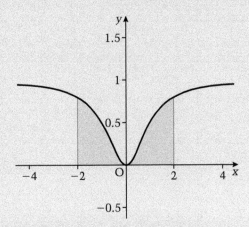

(a) The diagram shows the area between the curve $y = \dfrac{x^2}{1 + x^2}$, the x-axis and the ordinates $x = -2$ and $x = 2$

Use the trapezium rule with four intervals to find an approximate value for this area.

(b) Use the diagram to explain why it is difficult to judge whether your answer is an over-estimate or an under-estimate.

(c) Find the exact value of $\displaystyle\int_{-2}^2 \frac{x^2}{1 + x^2} \, dx$

2.1 Sequences

Learning outcomes

- To define a sequence
- To use a formula for the nth term or a recurrence relation to find a specific term of a sequence
- To define arithmetic and geometric progressions

Example

The nth term of a sequence is given by $u_n = \dfrac{n^2 + 1}{2n^2 - 3n}$

Find the value of u_4.

$$u_4 = \frac{4^2 + 1}{2(4^2) - 3(4)} = \frac{17}{20}$$

Sequences

A **sequence** is an ordered list of terms. There is a first term, a second term, and so on.

A sequence can have a finite number of terms or an infinite number of terms.

We denote the terms of a sequence as u_1, u_2, ..., u_n, ... where u_n is the nth term.

(The notation a_1, a_2, ..., a_n, ... is also used.)

When u_n is a function of n, we can use this to find a specific term.

For example, if the nth term of a sequence is given by $u_n = 2^n - 1$ then we can find a specific term by replacing n by the number of that term,

i.e. the first term is given by replacing n by 1, so $u_1 = 2^1 - 1 = 1$ similarly, $u_2 = 2^2 - 1 = 3$, $u_5 = 2^5 - 1 = 31$, $u_{10} = 2^{10} - 1 = 1023$, and so on.

Recurrence relations

Sometimes the terms of a sequence are related by a **recurrence relation**. This is an equation which connects the nth term to previous terms, for example, $u_n = 2u_{n-1} + 3$ or $u_n = u_{n-1} + u_{n-2}$. A recurrence relation on its own is not enough to define a sequence; we need to know the value of at least one term.

When $u_n = 2u_{n-1} + 3$, if we know the first term we can generate the sequence:

if $u_1 = 2$, the recurrence relation tells us that each term is twice the previous term plus 3,

so $u_2 = 4 + 3 = 7$, $u_3 = 2u_2 + 3 = 17$, and so on.

So the sequence is 2, 7, 17, 37, 77, ...

When $u_n = u_{n-1} + u_{n-2}$ we need to know the first two terms in order to generate the sequence.

If $u_1 = 2$ and $u_2 = 4$, then the recurrence relation tells us that each term is the sum of the two previous terms, so $u_3 = 4 + 2 = 6$, $u_4 = 6 + 4 = 10$, and so on.

So the sequence is 2, 4, 6, 10, 16, 26, ...

Any sequence where each term is the sum of the two previous terms (like the one above) is called a **Fibonacci sequence**.

Example

A sequence is given by $u_1 = 7$ and $u_{n+1} = 2u_n - 1$. Show that $u_n = 3(2^n) + 1$

$u_n = 3(2^n) + 1 \quad \Rightarrow \quad u_{n+1} = 3(2^{n+1}) + 1$

$\qquad\qquad\qquad\qquad = 2 \times 3(2^n) + 1 = 2(u_n - 1) + 1 = 2u_n - 1$

and $\quad u_n = 3(2^n) + 1 \quad \Rightarrow \quad u_1 = 7 \quad (n = 1)$

This verifies that the given formula for the nth term gives the first term and the recurrence relation.

Alternatively

$$u_{n+1} = 2u_n - 1 \Rightarrow u_{n+1} + 1 = 2u_n$$
$$\Rightarrow 3(2^{n+1}) + 1 + 1 = 2u_n \quad \text{Using } u_n + 1 = 3(2^{n+1}) + 1$$
$$\Rightarrow 6(2^n) + 2 = 2u_n$$
$$\Rightarrow u_n = 3(2^n) + 1$$

Example

The nth term of a sequence is given by $u_n = 5^n + 1$. Find u_{n+1} in terms of u_n.

$$u_n = 5^n + 1 \Rightarrow u_{n+1} = 5^{n+1} + 1$$
$$= 5(5^n) + 1$$
$$= 5(5^n + 1) - 4$$
$$= 5u_n - 4$$

Arithmetic progressions

An *arithmetic progression* (AP) is a sequence where each term differs by a constant from the previous term.

For example, 2, 5, 8, 11, 14, ... is an arithmetic progression as successive terms differ by 3.

A general AP whose first term is a and where the difference between successive terms is d (called the **common difference**) can be written as $a, a + d, a + 2d, a + 3d, ...$

The recurrence relation that gives an AP is $u_n = u_{n-1} + d$

We can see that the nth term is given by $u_n = a + (n - 1)d$

Geometric progressions

A *geometric progression* (GP) is a sequence where each term is a constant multiple of the previous term.

For example, 64, -32, 16, -8, 4, -2, ... is a GP as each term is $-\frac{1}{2}$ the previous term.

A general GP whose first term is a and where each term is the previous term multiplied by r (called the **common ratio**) can be written as $a, ar, ar^2, ar^3, ...$

The recurrence relation that gives a GP is $u_n = ru_{n-1}$

We can see that the nth term is given by $u_n = ar^{n-1}$

You need to be able to recognise an AP or a GP from a recurrence relation or from a formula for the nth term.

Exercise 2.1

1 State which of the following sequences are APs and which are GPs and in each case determine the 10th term.

(a) 5, 3, 1, -1, ...

(b) 1, $\frac{1}{2}$, $\frac{1}{4}$, $\frac{1}{8}$, ...

(c) 1, -1, 1, -1, 1, ...

(d) 1, $\frac{1}{2}$, 0, $-\frac{1}{2}$, -1, ...

2 A sequence is defined by $u_1 = 10$ and the recurrence relation $u_{n+1} = u_n - 3$. Find a formula for u_n in terms of n.

3 The nth term of a sequence is given by $u_n = 5^n - 4$. Find an equation giving u_{n+1} in terms of u_n.

4 The nth term of a sequence is given by $u_n = 3 \times 2^n - n$. Find the value of the 10th term.

Convergent sequences

Consider the sequence $1, 1\frac{1}{2}, 1\frac{1}{4}, 1\frac{1}{8}, 1\frac{1}{16}, \ldots$

The nth term of this sequence is given by $u_n = 1 + \dfrac{1}{2^n}$

Now as n increases, $\dfrac{1}{2^n} \to 0$, so $u_n \to 1$, i.e. $\lim\limits_{n \to \infty} (u_n) = 1$

So the terms of this sequence converge to the value 1, and the series is said to be **convergent**.

> **A sequence is convergent if the nth term is such that $\lim\limits_{n \to \infty} (u_n) = c$ where c is a finite constant.**

Divergent sequences

Consider the sequence $1, 3, 5, 7, 9, \ldots$

The nth term of this sequence is given by $u_n = 2n - 1$ and, as n increases, $2n - 1$ increases so $\lim\limits_{n \to \infty} (u_n) = \infty$. This sequence does not converge, the terms diverge and the sequence is said to be **divergent**.

> **A sequence that is not convergent is divergent.**

Example

Determine whether the sequence whose nth term is $\dfrac{4n^2 - 1}{5n^2 + 2n - 1}$ converges or diverges.

$$\frac{4n^2 - 1}{5n^2 + 2n - 1} = \frac{4 - \dfrac{1}{n^2}}{5 + \dfrac{2}{n} - \dfrac{1}{n^2}}$$

By dividing both numerator and denominator by n^2

$$\therefore \lim_{n \to \infty} \frac{4n^2 - 1}{5n^2 + 2n - 1} = \lim_{n \to \infty} \frac{4 - \dfrac{1}{n^2}}{5 + \dfrac{2}{n} - \dfrac{1}{n^2}} = \frac{4}{5}$$

Therefore the sequence converges.

Alternating sequences

When the terms in a sequence alternate between positive and negative, we have an **alternating** sequence.

For example, $1, -1, 1, -1, 1, \ldots$ and $0.5, -0.05, 0.005, -0.0005, \ldots$ are alternating sequences.

An alternating sequence may be convergent or divergent.

The nth term of $1, -1, 1, -1, 1, \ldots$ is given by $u_n = (-1)^{n+1}$ and $\lim\limits_{n \to \infty} (u_n)$ does not exist, so this sequence is divergent.

The nth term of $0.5, -0.05, 0.005, -0.0005, \ldots$ is given by

$u_n = \frac{1}{2}(10^{1-n})(-1)^{n+1}$ and $\lim\limits_{n \to \infty} (u_n) = 0$ so this sequence is convergent.

Note that a negative number to a power involving a multiple of n will alternate between positive and negative values.

Periodic sequences

When the terms of a sequence form a repeating pattern, the sequence is called **periodic**. For example, 1, 2, 3, 1, 2, 3, 1, 2, 3, ... is periodic.

A periodic sequence may also be alternating, for example, 1, −1, 1, −1, 1, ...

Periodic sequences are not convergent.

Oscillating sequences

The terms in an **oscillating** sequence move between higher and lower values.

Examples of oscillating sequences are

(a) 1, −1, 1, −1, ... **(b)** 1, 2, 3, 1, 2, 3, ...
(c) 1, 0, 2, 0, 3, 0, ... **(d)** 5, −5, 6, −6, 7, −7, ...

An oscillating sequence may be an alternating sequence as in **(a)** and **(d)**, or an oscillating sequence may be a periodic sequence as in **(a)** and **(b)**.

Oscillating sequences are not convergent.

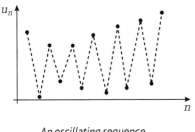

An oscillating sequence

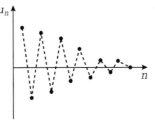

An alternating sequence (also convergent)

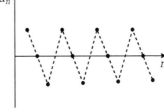

A periodic sequence (also oscillating)

Note that the nth term of an arithmetic progression is $u_n = a + (n - 1)d$ so $\lim_{n \to \infty} (u_n) = \infty$

Therefore all arithmetic progressions are divergent.

The nth term of a geometric progression is $u_n = ar^{n-1}$ and $\lim_{n \to \infty} (u_n)$ depends on the value of r.

If $-1 < r < 1$, $r^{n-1} \to 0$ as $n \to \infty$ so $\lim_{n \to \infty} (u_n) = 0$ and the sequence is convergent.

Example

Determine whether the sequence whose nth term is given by $u_n = 5 - \left(-\frac{1}{3}\right)^{n+1}$ is alternating, periodic, oscillating or none of these.

$\lim_{n \to \infty} (u_n) = 5$, so the sequence converges and so is neither periodic nor oscillating.

$\left(-\frac{1}{3}\right)^{n+1}$ alternates in sign, but $\left|\left(-\frac{1}{3}\right)^{n+1}\right| < 1$ therefore $5 - \left(-\frac{1}{3}\right)^{n+1}$ is always positive.

Therefore the sequence is not alternating.

Exercise 2.2

1 Determine which of the following sequences, whose nth term is given, converges.

(a) $\dfrac{n + 1}{n^2 + 1}$ **(b)** $\dfrac{2n^2 + 1}{n^2 + 1}$

(c) $\dfrac{n^3 + 1}{n^2 + 1}$ **(d)** $(-1)^n$

2 Determine whether each of the following sequences is alternating, periodic, oscillating or none of these.

(a) $u_1 = -1$, $u_2 = 1$ and $u_{n+2} = u_{n+1} + 2u_n$

(b) $u_n = \cos n\pi$ **(c)** $u_n = \sin\dfrac{n\pi}{2}$

2.3 Number series

Learning outcomes

- To define a number series
- To introduce the $\sum$ notation
- To use the sum of the first n terms of a series to find the sum to infinity of the series
- To define convergence and divergence of series

You need to know

- The meaning of an arithmetic progression and a geometric progression
- The general term of an arithmetic sequence and of a geometric sequence

☑ Exam tip

When you are finding a general term of a sequence or a series, look for a relationship between the term number, r, and the numbers in the term. Common relationships are multiples r, of $r \pm k$, multiples of $r^2 \pm k$, where k is a constant.

Series

A series is the sum of the terms of a sequence.

For example, $1 + 2 + 3 + 4 + \ldots$ is a series.

When the terms are real numbers the series is called a **number series**.

We use u_r to denote a general term of a series.

Example

Find the rth term of the series $\dfrac{1}{(2)(3)} + \dfrac{2}{(3)(4)} + \dfrac{3}{(4)(5)} + \ldots$

The numerator of each term is equal to the term number, r, and the denominator is the product of $r + 1$ and $r + 2$

Therefore $u_r = \dfrac{r}{(r + 1)(r + 2)}$

Check to see that the answer does give the first 3 terms.

The sum of the first n terms of a series

The nth term of the series $1 + 2 + 3 + 4 + \ldots$ is n.

The sum of the first n terms is $1 + 2 + 3 + \ldots + (n - 1) + n$

We can write this more briefly using $\sum$ to mean 'the sum of'.

Taking the rth term as a general term (i.e. any term between the first and nth term),

then $\displaystyle\sum_{r=1}^{r=n} r$ means the sum of all the values of r from $r = 1$ to $r = n$

i.e. $\displaystyle\sum_{r=1}^{r=n} r = 1 + 2 + 3 + \ldots + (n - 1) + n$

Similarly $\displaystyle\sum_{r=1}^{r=n} \dfrac{1}{r + 1}$

means $\dfrac{1}{2} + \dfrac{1}{3} + \dfrac{1}{4} + \ldots + \dfrac{1}{n + 1}$

The sum of the first n terms of an arithmetic progression

Any AP has the form $a,\ a + d,\ a + 2d,\ a + 3d,\ \ldots$ where the rth term is $a + (r - 1)d$

The sum of the first n terms is

$$\sum_{r=1}^{r=n} (a + (r - 1)d) = a + (a + d) + (a + 2d) + \ldots + (a + (n - 1)d)$$

Using $S_n = \displaystyle\sum_{r=1}^{r=n} (a + (r - 1)d)$ we have

$$S_n = \quad a \quad + \quad (a + d) \quad + \quad (a + 2d) \quad + \dots + (a + (n - 1)d) \qquad [1]$$

and (writing the right-hand side in reverse order)

$$S_n = (a + (n - 1)d) + (a + (n - 2)d) + (a + (n - 3)d) + \dots + \quad a \qquad [2]$$

Adding [1] and [2] gives $2S_n = n(2a + (n - 1)d)$

Therefore

$$S_n = \frac{n}{2}(2a + (n - 1)d$$

You may quote this formula unless you are asked to derive it.

An alternative version of the formula above is $S_n = \frac{n}{2}(a + l)$ where l is the last term. This version is derived from [2] where $S_n = \frac{n}{2}(a + a(n - 1)d)$

For example, the terms of the series $1 + 3 + 5 + 7 + \dots$ are an AP, where $a = 1$ and $d = 2$

The sum of the first n terms is given by $S_n = \frac{n}{2}(2 + 2(n - l)) = n^2$

The sum of the first n terms of a geometric progression

Any GP has the form $a, ar, ar^2, ar^3, \dots, ar^{n-1}, \dots$

The sum of the first n terms is given by

$$S_n = a + ar + ar^2 + ar^3 + \dots + ar^{n-1} \qquad [1]$$

Now $\quad rS_n = \quad ar + ar^2 + ar^3 + \dots + ar^{n-1} + ar^n \qquad [2]$

[1] − [2] gives $\quad S_n(1 - r) = a - ar^n$

Therefore

$$S_n = \frac{a(1 - r^n)}{1 - r}$$

You may also quote this formula unless you are asked to derive it.

Example

Find $\displaystyle\sum_{r=1}^{r=m+1} 5\left(\frac{1}{3}\right)^r$

$\displaystyle\sum_{r=1}^{r=m+1} 5\left(\frac{1}{3}\right)^r$ is recognised as the sum of the first $m + 1$ terms of a

GP, with first term $\frac{5}{3}$ and common ratio $\frac{1}{3}$

Therefore $\displaystyle\sum_{r=1}^{r=m+1} 5\left(\frac{1}{3}\right)^r = \dfrac{\frac{5}{3}\left(1 - \left(\frac{1}{3}\right)^{m+1}\right)}{1 - \frac{1}{3}}$

$$= \frac{5}{2}\left(1 - \left(\tfrac{1}{3}\right)^{m+1}\right)$$

☑ *Exam tip*

If you do not recognise the form of a series, write out the first few terms:

in this example using $\displaystyle\sum_{r=1}^{r=m+1} 5\left(\frac{1}{3}\right)^r$

with $r = 1, 2, 3, \dots$ gives

$\frac{5}{3} + \frac{5}{3^2} + \frac{5}{3^3} + \dots$

Example

Given $\displaystyle\sum_{r=1}^{r=n} u_r = n(2 - n^2)$ find u_n.

$$\sum_{r=1}^{r=n} u_r = u_1 + u_2 + \ldots + u_{n-1} + u_n \quad \text{and} \quad \sum_{r=1}^{r=n-1} u_r = u_1 + u_2 + \ldots + u_{n-1}$$

Therefore $\displaystyle u_n = \sum_{r=1}^{r=n} u_r - \sum_{r=1}^{r=n-1} u_r$

$$= n(2 - n^2) - (n-1)(2 - (n-1)^2)$$
$$= 1 + 3n - 3n^2$$

Exercise 2.3a

1 Find the rth term of the series

    (a) $\dfrac{1}{2} + \dfrac{2}{5} + \dfrac{3}{10} + \dfrac{4}{17} + \ldots$

    (b) $\dfrac{1}{(2)(4)} + \dfrac{1}{(3)(7)} + \dfrac{1}{(4)(10)} + \dfrac{1}{(5)(13)} + \ldots$

2 Find the sum of the first n terms of the series

$$1 + \frac{1}{2} + 0 - \frac{1}{2} - 1 - \frac{3}{2} - \ldots$$

3 Evaluate

    (a) $\displaystyle\sum_{r=1}^{r=5} 3\left(\frac{1}{2}\right)^r$

    (b) $\displaystyle\sum_{r=1}^{r=10} 3\left(\frac{1}{2}\right)^r$

    Hence find $\displaystyle\sum_{r=6}^{r=10} 3\left(\frac{1}{2}\right)^r$

4 Given $\displaystyle\sum_{r=1}^{r=n} u_r = \frac{3-n}{3+n}$, find u_n in terms of n.

5 Given $\displaystyle\sum_{r=1}^{r=n} u_r = \frac{n}{n+1}$, find $\displaystyle\sum_{r=10}^{r=20} u_r$

The sum to infinity of a series

The sum of the first n terms of a series is given by $\displaystyle\sum_{r=1}^{r=n} u_r = 1 + \frac{1}{n}$

As $n \to \infty$, $1 + \dfrac{1}{n} \to 1$ so the sum of the terms of this series converges to 1.

This is called the **sum to infinity** of the series.

> A series is convergent when the sum to infinity (i.e. $\displaystyle\lim_{n\to\infty}\left(\sum_{r=1}^{r=n} u_r\right)$)
> is a finite constant.

If $\displaystyle\sum_{r=1}^{r=n} u_r = n^2 + 1$, then as $n \to \infty$, $n^2 + 1 \to \infty$ so $\displaystyle\lim_{n\to\infty}\left(\sum_{r=1}^{r=n} u_r\right)$
is not a finite constant and the series diverges.

A series that does not converge is called a divergent series.

Clearly any series, $u_1 + u_2 + u_3 + \dots$, where the terms do not approach zero as n increases, will diverge as the sum of the first n terms will continue to increase.

Therefore a necessary (but not sufficient) condition for a series to converge is that the nth term approaches zero as n approaches infinity.

Arithmetic progressions and geometric progressions

The sum of the first n terms of an AP is $\frac{n}{2}(2a + (n-1)d)$ and it is clear that this sum diverges as $n \to \infty$

Therefore the sum of an AP always diverges.

The sum of the first n terms of a GP is $\frac{a(1 - r^n)}{1 - r}$ where a is the first term and r is the common ratio.

Whether $\lim\limits_{n \to \infty} \left(\frac{a(1 - r^n)}{1 - r} \right)$ is a constant depends on the value of r:

If $|r| > 1$, $r^n \to \infty$ as $n \to \infty$ so $\frac{a(1 - r^n)}{1 - r} \to \infty$ and the series diverges.

If $r = 1$, $1 - r = 0$ so $\frac{a(1 - r^n)}{1 - r}$ is meaningless.

If $|r| < 1$, $r^n \to 0$ as $n \to \infty$ so $\frac{a(1 - r^n)}{1 - r} \to \frac{a}{1 - r}$ and the series converges.

Therefore, provided that $|r| < 1$, the sum to infinity of a GP is $\dfrac{a}{1 - r}$

Example
Show that the series $\frac{1}{3} + \frac{1}{3^3} + \frac{1}{3^5} + \frac{1}{3^7} + \dots$ is geometric and find the sum to infinity of this series.

$u_1 = \frac{1}{3}$, $u_2 = u_1 \times \left(\frac{1}{3^2}\right)$, $u_3 = u_2 \times \left(\frac{1}{3^2}\right)$ and so on. Therefore each term is $\frac{1}{3^2}$ times the previous term, so the series is geometric with first term $\frac{1}{3}$ and common ratio $\frac{1}{3^2}$ (<1).

So the sum to infinity is $\dfrac{\frac{1}{3}}{1 - \frac{1}{3^2}} = \dfrac{3}{8}$

Exercise 2.3b

1 S_n is the sum of the first n terms of a series. Determine whether the series is convergent, and if it is, give the sum to infinity when

(a) $S_n = \dfrac{2n}{n + 1}$ (b) $S_n = \dfrac{n^2}{n + 1}$ (c) $S_n = 2^{-n}$

2 (a) Show that the series $\frac{1}{2} - \frac{1}{6} + \frac{1}{18} - \frac{1}{54} + \dots$ is geometric.

   (b) Find the sum of the first n terms of the series in (a) and hence find the sum to infinity.

Finding the sum of the first n terms of a series

We have found the sum of the first n terms of a series whose terms are in arithmetic progression and a series whose terms are in geometric progression. There is no general method that will give the sum of the first n terms of any series, but there are methods that work for some types of series.

Method of differences

This method works with a series whose general term can be expressed as $f(r + 1) - f(r)$, because most of the terms cancel when they are listed.

Consider the series $\dfrac{1}{1 \times 2} + \dfrac{1}{2 \times 3} + \dfrac{1}{3 \times 4} + \ldots + \dfrac{1}{r(r + 1)} + \ldots$

So $u_r = \dfrac{1}{r(r + 1)}$ and we use partial fractions to express this as two separate fractions:

$$\frac{1}{r(r + 1)} = \frac{A}{r} + \frac{B}{r + 1} \quad \Rightarrow \quad 1 \equiv A(r + 1) + Br$$

$$r = 0 \quad \Rightarrow \quad A = 1 \quad \text{and} \quad r = -1 \quad \Rightarrow \quad B = -1$$

$$\therefore \quad \frac{1}{r(r + 1)} = \frac{1}{r} - \frac{1}{r + 1}$$

Hence $\displaystyle\sum_{r=1}^{r=n} u_r = \sum_{r=1}^{r=n} \left(\frac{1}{r} - \frac{1}{r + 1}\right)$

We now list the terms vertically (this makes it easier to see the terms that cancel):

✅ Exam tip

You need to list enough terms at the start and at the end so that you can clearly see the pattern of cancelling.

$$\sum_{r=1}^{r=n} \left(\frac{1}{r} - \frac{1}{r + 1}\right) = \frac{1}{1} - \frac{1}{2}$$

$$+ \frac{1}{2} - \frac{1}{3}$$

$$+ \frac{1}{3} - \frac{1}{4}$$

$$+ \frac{1}{4} - \frac{1}{5} \ldots$$

$$+ \ldots - \ldots$$

$$+ \frac{1}{n - 1} - \frac{1}{n}$$

$$+ \frac{1}{n} - \frac{1}{n + 1} = 1 - \frac{1}{n + 1} = \frac{n}{n + 1}$$

$$\therefore \quad \sum_{r=1}^{n} \frac{1}{r(r + 1)} = \frac{n}{n + 1}$$

Example

(a) Express $\dfrac{1}{(r-1)(r-2)}$ in partial fractions.

(b) Hence find $\displaystyle\sum_{r=3}^{r=n} \dfrac{1}{(r-1)(r-2)}$

(c) Deduce the sum to infinity of the series

$$\dfrac{1}{2\times 1} + \dfrac{1}{3\times 2} + \dfrac{1}{4\times 3} + \dots$$

(a) $\dfrac{1}{(r-1)(r-2)} = \dfrac{A}{r-1} + \dfrac{B}{r-2} \Rightarrow 1 \equiv A(r-2) + B(r-1)$

$r = 1 \Rightarrow A = -1$ and $r = 2 \Rightarrow B = 1$

$\therefore \dfrac{1}{(r-1)(r-2)} = -\dfrac{1}{r-1} + \dfrac{1}{r-2}$

(b) Note that the first term of this series is given by $r = 3$,

so $\dfrac{1}{(r-1)(r-2)}$ is the $(r-2)$th term.

$$\sum_{r=3}^{r=n} \dfrac{1}{(r-1)(r-2)} = \sum_{r=3}^{r=n}\left(-\dfrac{1}{r-1} + \dfrac{1}{r-2}\right) = -\dfrac{1}{2} + \dfrac{1}{1}$$

$$-\dfrac{1}{3} + \dfrac{1}{2}$$
$$-\dfrac{1}{4} + \dfrac{1}{3}$$
$$-\dots + \dots$$
$$-\dfrac{1}{n-2} + \dfrac{1}{n-3}$$
$$-\dfrac{1}{n-1} + \dfrac{1}{n-2}$$
$$= 1 - \dfrac{1}{n-1} = \dfrac{n-2}{n-1}$$

(c) Sum to infinity $= \displaystyle\lim_{n\to\infty}$ (the sum of the terms up to $r = n$)

$$= \lim_{n\to\infty}\left(1 - \dfrac{1}{n-1}\right) = 1$$

Exercise 2.4

1 (a) Express $\dfrac{1}{(r-1)(r+1)}$ in partial fractions.

(b) Hence find $\displaystyle\sum_{r=2}^{r=n}\dfrac{1}{(r-1)(r+1)}$

(c) Deduce the sum to infinity of the series

$$\dfrac{1}{1\times 3} + \dfrac{1}{2\times 4} + \dfrac{1}{3\times 5} + \dots$$

2 Express $\dfrac{1}{r(r+1)(r+2)}$ in partial fractions and hence find

$$\sum_{r=1}^{r=n}\left(\dfrac{1}{r(r+1)(r+2)}\right)$$

Learning outcomes

- To use proof by induction to prove properties of sequences and series

You need to know

- How to use proof by induction

A formula for the nth term of a sequence

When a sequence is defined by a recurrence relation, we may be able to deduce a formula for the nth term that works for the first few terms, but we need to prove that it works for all the terms. We can do this using proof by induction.

> ### Example
>
> A sequence of positive integers, $\{U_n\}$, is defined by $U_1 = 1$ and $3U_{n+1} = 2U_n - 1$
>
> Prove by mathematical induction that $U_n = 3\left(\frac{2}{3}\right)^n - 1$
>
> Let P_n be the statement $U_n = 3\left(\frac{2}{3}\right)^n - 1$
>
> Now P_1 is $U_1 = 3\left(\frac{2}{3}\right)^1 - 1 = 1$, which is true.
>
> Assume that P_n is true when $n = k$, i.e. that $U_k = 3\left(\frac{2}{3}\right)^k - 1$
>
> Using the recurrence relation gives
>
> $$3U_{k+1} = 2\left(3\left(\frac{2}{3}\right)^k - 1\right) - 1$$
>
> $$\Rightarrow \quad U_{k+1} = \frac{2}{3}\left(3\left(\frac{2}{3}\right)^k - 1\right) - \frac{1}{3} = 3\left(\frac{2}{3}\right)^{k+1} - 1$$
>
> Therefore if P_k is true, P_{k+1} is also true.
>
> As P_k is true when $k = 1$, then it is true when $k = 2, 3, 4, \ldots, n$
>
> Therefore $U_n = 3\left(\frac{2}{3}\right)^n - 1$ is true for all $n \in \mathbb{N}$

A formula for the sum of the first n terms of a series

It is not always possible to find the sum of the first n terms of a given series. We may be able to deduce a formula that works for the first few terms, but we need to prove that it works for all the terms. We can do this using proof by induction.

Consider, for example, the series $1^2 + 2^2 + 3^2 + 4^2 + \ldots + n^2 + \ldots$

Now when $r = 1, 2, 3, 4, \ldots$ $\sum_{r=1}^{n} r^2$ gives the sequence 1, 5, 14, 30, ...

From this we may be able to deduce that $\sum_{r=1}^{n} r^2 = \frac{n}{6}(n+1)(2n+1)$ is true for $n = 1, 2, 3, 4$

Example

Prove by mathematical induction that $\sum_{r=1}^{n} r^2 = \frac{n}{6}(n+1)(2n+1)$ for all $n \in \mathbb{N}$

Let P_n be the statement $\sum_{r=1}^{n} r^2 = \frac{n}{6}(n+1)(2n+1)$

When $n = 1$, $P_1 = \frac{1}{6}(2)(3) = 1 = 1^2$, i.e. P_1 is true.

Assume that P_n is true when $n = k$, i.e. $\qquad P_k = \sum_{r=1}^{k} r^2 = \frac{k}{6}(k+1)(2k+1)$ [1]

then adding the next term of the series gives $P_{k+1} = \sum_{r=1}^{k+1} r^2 = \frac{k}{6}(k+1)(2k+1) + (k+1)^2$

We now aim to simplify the right-hand side so that it becomes [1] with $k+1$ replacing k.

$$\frac{k}{6}(k+1)(2k+1) + (k+1)^2 = (k+1)\left(\frac{k}{6}(2k+1) + (k+1)\right) = \frac{(k+1)}{6}(k(2k+1) + 6(k+1))$$

$$= \frac{(k+1)}{6}(2k^2 + 7k + 6)$$

$$= \frac{(k+1)}{6}(k+2)(2k+3)$$

$$= \frac{(k+1)}{6}[(k+1)+1][2(k+1)+1]$$

Therefore if P_k is true, P_{k+1} is also true.

As P_k is true when $k = 1$, then it is true when $k = 2, 3, 4, \ldots, n$

Therefore $\sum_{r=1}^{n} r^2 = \frac{n}{6}(n+1)(2n+1)$ is true for all $n \in \mathbb{N}$

There are some number series whose sums are worth remembering. These are:

the sum of the first n natural numbers: $\sum_{r=1}^{n} r = \frac{n}{2}(n+1)$

(This is the sum of the terms of an AP so can be verified using the formula derived in Topic 2.3.)

the sum of the squares of the first n natural numbers:

$$\sum_{r=1}^{n} r^2 = \frac{n}{6}(n+1)(2n+1)$$

(This is proved above.)

the sum of the cubes of the first n natural numbers:

$$\sum_{r=1}^{n} r^3 = \frac{n^2}{4}(n+1)^2$$

(This can be proved by induction and is part of question 1 in Exercise 2.5 below.)

These results can be used to find the sums of series whose general term is the sum or difference of ar, ar^2 and/or ar^3.

Example

Find $\sum_{r=1}^{n} r(2r+1)$

$$\sum_{r=1}^{n} r(2r+1) = \sum_{r=1}^{n}(2r^2 + r)$$

$$= 2\sum_{r=1}^{n} r^2 + \sum_{r=1}^{n} r$$

$$= 2\left(\frac{n}{6}(n+1)(2n+1)\right)$$

$$+ \left(\frac{n}{2}(n+1)\right)$$

$$= \frac{n}{6}(n+1)(4n+5)$$

Exercise 2.5

1 Prove by induction that

(a) $\sum_{r=2}^{n} \frac{1}{r(r-1)} = \frac{n-1}{n}$ $\qquad$ (b) $\sum_{r=1}^{n} r^3 = \frac{n^2}{4}(n+1)^2$

2 (a) Find the rth term of the series $1(4) + 2(7) + 3(10) + 4(13)$

(b) Prove by induction that the sum of the first n terms of this series is $n(n+1)^2$

Learning outcomes

- To define a power series
- To introduce the factorial notation
- To derive and use Maclaurin's theorem to expand functions as a power series

You need to know

- How to differentiate simple functions
- How to differentiate products of functions
- The values of the trig ratios for multiples (including fractional) of π

Power series

A series whose terms involve increasing or decreasing integral powers of a variable is called a **power series**.

For example $2 + 3x + 4x^2 + 5x^3 + ...$ and $x^n - x^{n-1} + x^{n-2} - ...$ are power series.

The factorial notation

There are several occasions when products such as $1 \times 2 \times 3 \times 4 \times 5 \times ... \times 40$ occur.

There is a shorthand notation for products such as these.

We denote $1 \times 2 \times 3$ by 3! (called 3 **factorial**).

6! means the product of all the integers from 1 to 6 inclusive and $n!$ means the product of all the integers from 1 to n inclusive,

i.e. $n! = (1)(2)(3) ... (n-2)(n-1)(n)$

Example

Evaluate $\dfrac{20!}{17!3!}$

20! is the product of the integers from 1 to 20 and 17! is the product of the integers from 1 to 17, so we can cancel this product.

$\therefore \quad \dfrac{20!}{17!3!} = \dfrac{18 \times 19 \times 20}{3 \times 2 \times 1} = 3 \times 19 \times 20 = 1140$

Exercise 2.6a

Evaluate

1   4!   **2**   5!   **3** $\dfrac{5!}{3!}$ **4** $\dfrac{9!}{3!6!}$ **5** $\dfrac{3 \times 4}{5!}$

Maclaurin's theorem

If we assume that a function of x, $f(x)$, can be expanded as a series of ascending powers of x and that this series can be differentiated term by term, then

$$f(x) = a_0 + a_1x + a_2x^2 + a_3x^3 + a_4x^4 + ... + a_rx^r + ... \qquad [1]$$

where $a_0, a_1, a_2, ...$ are constants.

Substituting 0 for x in [1] gives $f(0) = a_0$,   i.e. $a_0 = f(0)$

Differentiating [1] with respect to x gives

$$f'(x) = a_1 + 2a_2x + 3a_3x^2 + 4a_4x^3 + 5a_5x^4 + ... \qquad [2]$$

Substituting 0 for x in [2] gives $f'(0) = a_1$,   i.e. $a_1 = f'(0)$

Differentiating [2] with respect to x gives

$$f''(x) = 2a_2 + (2)(3)a_3 x + (3)(4)a_4 x^2 + (4)(5)a_5 x^3 + \ldots \qquad [3]$$

Substituting 0 for x in [3] gives $f''(0) = 2a_2$,   i.e. $a_2 = \dfrac{f''(0)}{2!}$

Differentiating [3] with respect to x gives

$$f'''(x) = (2)(3)a_3 + (2)(3)(4)a_4 x + (3)(4)(5)a_5 x^2 + \ldots \qquad [4]$$

Substituting 0 for x in [4] gives $f'''(0) = (2)(3)a_3$,   i.e. $a_3 = \dfrac{f'''(0)}{3!}$

After differentiating r times we get

$$f^r(x) = (2)(3)(4)\ldots(r-1)(r)a_r + (2)(3)\ldots(r+1)a_{r+1}x + \ldots$$

Substituting 0 for x gives $f^r(0) = r!a_r$ i.e. $a_r = \dfrac{f^r(0)}{r!}$

Substituting these values for $a_1, \ldots$ in [1] gives

$$f(x) = f(0) + f'(0)x + \frac{f''(0)}{2!}x^2 + \frac{f'''(0)}{3!}x^3 + \ldots + \frac{f^r(0)}{r!}x^r + \ldots$$

$$= \sum_{n=0}^{\infty} \frac{f^n(0)x^n}{n!}$$

This is called *Maclaurin's theorem* and you need to learn it.

The series can be found if the nth derivative of $f(x)$ exists when $x = 0$ for all values of n. For the series expansion to equal $f(x)$, the series must converge to $f(x)$.

Some series converge to $f(x)$ for all values of x and some converge for a limited range of values of x. In the following examples, the values of x for which the series converges is given but without proof.

Example

Use Maclaurin's theorem to find the power series expansion of $f(x) = e^x$

Using $f(x) = f(0) + f'(0)x + \dfrac{f''(0)}{2!}x^2 + \dfrac{f'''(0)}{3!}x^3 + \ldots + \dfrac{f^r(0)}{r!}x^r + \ldots$

gives $f(x) = e^x$ so $f(0) = e^0 = 1$

           $f'(x) = e^x$ so $f'(0) = e^0 = 1$

           $f''(x) = e^x$ so $f''(0) = e^0 = 1$

           $f'''(x) = e^x$ so $f'''(0) = e^0 = 1$

        $\ldots$

           $f^r(x) = e^x$ so $f^r(0) = e^0 = 1$

Therefore $e^x = 1 + x + \dfrac{x^2}{2!} + \dfrac{x^3}{3!} + \dfrac{x^4}{4!} + \ldots + \dfrac{x^r}{r!} + \ldots$

This series converges for all values of x.

Example

Use Maclaurin's theorem to find the power series expansion of $f(x) = \cos x$

Using $f(x) = f(0) + f'(0)x + \dfrac{f''(0)}{2!}x^2 + \dfrac{f'''(0)}{3!}x^3 + \ldots + \dfrac{f^r(0)}{r!}x^r + \ldots$

gives

$$
\begin{aligned}
f(x) &= \cos x &\text{so} \quad f(0) &= \cos 0 = 1 \\
f'(x) &= -\sin x &\text{so} \quad f'(0) &= -\sin 0 = 0 \\
f''(x) &= -\cos x &\text{so} \quad f''(0) &= -\cos 0 = -1 \\
f'''(x) &= \sin x &\text{so} \quad f'''(0) &= \sin 0 = 0 \\
f''''(x) &= \cos x &\text{so} \quad f''''(0) &= \cos 0 = 1
\end{aligned}
$$

Therefore $\cos x = 1 + (0)x - \dfrac{x^2}{2!} + \dfrac{(0)x^3}{3!} + \dfrac{x^4}{4!} + \ldots$

We can see that values cycle from 1 to 0 to -1 to 0 to 1 again and so the series involves only even powers of x. Therefore the general term has the form $\pm \dfrac{x^{2r}}{(2r)!}$; when r is odd the term is negative, and when r is even the term is positive. We can show this using $(-1)^r$,

i.e. $\quad \cos x = 1 - \dfrac{x^2}{2!} + \dfrac{x^4}{4!} - \ldots + (-1)^r \dfrac{x^{2r}}{2r!} + \ldots$

This series converges for all values of x.

Example

Use Maclaurin's theorem to find the series expansion of $f(x) = \ln(1 + x)$

Using $f(x) = f(0) + f'(0)x + \dfrac{f''(0)}{2!}x^2 + \dfrac{f'''(0)}{3!}x^3 + \ldots + \dfrac{f^r(0)}{r!}x^r + \ldots$

gives

$$
\begin{aligned}
f(x) &= \ln(1 + x) &\text{so} \quad f(0) &= \ln 1 = 0 \\
f'(x) &= \frac{1}{1+x} &\text{so} \quad f'(0) &= 1 \\
f''(x) &= -\frac{1}{(1+x)^2} &\text{so} \quad f''(0) &= -1 \\
f'''(x) &= +\frac{2}{(1+x)^3} &\text{so} \quad f'''(0) &= 2 \\
f''''(x) &= -\frac{2 \times 3}{(1+x)^4} &\text{so} \quad f''''(0) &= -2 \times 3 = -3!
\end{aligned}
$$

Therefore $\ln(1 + x) = 0 + x - \dfrac{x^2}{2!} + \dfrac{2x^3}{3!} - \dfrac{3!x^4}{4!} + \ldots$ and the general term has the form

$$\pm \frac{(r-1)!x^r}{r!} = \pm \frac{x^r}{r}$$

This term is positive when r is negative and vice-versa, which we can show using $(-1)^{r+1}$

$$\Rightarrow \quad \ln(1 + x) = x - \frac{x^2}{2} + \frac{x^3}{3} - \frac{x^4}{4} + \ldots + (-1)^{r+1}\frac{x^r}{r} + \ldots$$

This series converges for $-1 < x \leqslant 1$

Note that it is not possible to use Maclaurin's theorem to expand $f(x) = \ln x$ because $f(0) = \ln 0$ and $\ln 0$ is undefined.

Standard expansions

These are the series you are expected to know:

$$e^x = 1 + x + \frac{x^2}{2!} + \frac{x^3}{3!} + \frac{x^4}{4!} + \ldots + \frac{x^r}{r!} + \ldots \qquad \text{for all values of } x$$

$$\cos x = 1 - \frac{x^2}{2!} + \frac{x^4}{4!} - \ldots + (-1)^r \frac{x^{2r}}{2r!} + \ldots \qquad \text{for all values of } x$$

$$\sin x = x - \frac{x^3}{3!} + \frac{x^5}{5!} + \ldots + (-1)^r \frac{x^{2r+1}}{(2r+1)!} + \ldots \qquad \text{for all values of } x$$

$$\ln(1 + x) = x - \frac{x^2}{2} + \frac{x^3}{3} - \frac{x^4}{4} + \ldots + (-1)^{r+1} \frac{x^r}{r} + \ldots \qquad \text{for } -1 < x \leqslant 1$$

$$\ln(1 - x) = -x - \frac{x^2}{2} - \frac{x^3}{3} - \frac{x^4}{4} - \ldots - (-1)^{r+1} \frac{x^r}{r} + \ldots \qquad \text{for } -1 \leqslant x < 1$$

Example

Expand $e^x \sin 2x$ as a power series as far as the term in x^3.

Using the standard expansions for e^x and $\sin x$ as far as the term in x^3 gives

$$e^x = 1 + x + \frac{x^2}{2!} + \frac{x^3}{3!} + \ldots \quad \text{and} \quad \sin 2x = (2x) - \frac{(2x)^3}{3!} + \ldots \qquad \text{Replacing } x \text{ with } 2x$$

$$\therefore \quad e^x \sin 2x = \left(1 + x + \frac{x^2}{2!} + \frac{x^3}{3!} + \ldots\right)\left(2x - \frac{4x^3}{3} + \ldots\right)$$

Multiplying the brackets and ignoring any terms involving powers of x greater than 3 gives

$$e^x \sin 2x = 2x - \frac{x^3}{3} + 2x^2 + \ldots = 2x + 2x^2 - \frac{x^3}{3} + \ldots$$

The series found so far have been infinite, but some series terminate.
For example, using Maclaurin's theorem to expand $(1 + x)^4$ gives

$$f(x) = f(0) + f'(0)x + \frac{f''(0)}{2!}x^2 + \frac{f'''(0)}{3!}x^3 + \ldots + \frac{f^r(0)}{r!}x^r + \ldots$$

$\therefore \quad f(x) = (1 + x)^4, \qquad f'(x) = 4(1 + x)^3, \qquad f''(x) = 12(1 + x)^2, \qquad f'''(x) = 24(1 + x), \qquad f''''(x) = 24$

so $\quad f(0) = 1, \qquad\qquad f'(0) = 4, \qquad\qquad f''(0) = 12, \qquad\qquad f'''(0) = 24, \qquad\qquad f''''(0) = 24$

All further differentials are 0, so the series terminates.

$$\therefore \quad (1 + x)^4 = 1 + 4x + \frac{12}{2!}x^2 + \frac{24}{3!}x^3 + \frac{24}{4!}x^4$$

i.e. $(1 + x)^4 = 1 + 4x + 6x^2 + 4x^3 + x^4$

Note that there are easier ways to expand functions of the form $(1 + x)^n$ which we will look at later in this section.

Exercise 2.6b

Use Maclaurin's theorem to expand each of the following functions as far as the term in x^4 and give the range of values of x for which they are valid.

1 $f(x) = e^{-x}$

2 $\tan 2x$

3 $\ln(1 - 3x)$

4 $e^x \cos x$

5 (a) Use Maclaurin's theorem to show that
$$(1 - x)^{-1} = 1 + x + x^2 + x^3 + x^4 + x^5 + \ldots$$

(b) Write down u_n and u_{n+1} where u_n and u_{n+1} are the nth and $(n+1)$th terms respectively of this series.
Hence find a recurrence relation between u_n and u_{n+1}

(c) Use the recurrence relation to show that the series is geometric and hence verify that the series converges to $(1 - x)^{-1}$, stating the range of values for which this is true.

Learning outcomes

- To prove Euler's formula
- To use Maclaurin's theorem to expand further functions and find approximations

You need to know

- How to evaluate powers of i (i.e. $\sqrt{-1}$)
- The standard Maclaurin series
- The values of the trig ratios for multiples (including fractional) of π
- The meaning of a quadratic function

Euler's formula

We introduced the formula $e^{i\theta} = \cos\theta + i\sin\theta$ in Topic 1.7.

We can now use the Maclaurin expansions of $\cos\theta$ and $\sin\theta$ to prove it:

$$\cos\theta = 1 - \frac{\theta^2}{2!} + \frac{\theta^4}{4!} - \ldots + (-1)^r \frac{\theta^{2r}}{2r!} + \ldots$$

and $\sin\theta = \theta - \frac{\theta^3}{3!} + \frac{\theta^5}{5!} + \ldots + (-1)^r \frac{\theta^{2r+1}}{(2r+1)!} + \ldots$

$\therefore \quad \cos\theta + i\sin\theta = 1 + i\theta - \frac{\theta^2}{2!} - \frac{i\theta^3}{3!} + \frac{\theta^4}{4!} + \frac{i\theta^5}{5!} + \ldots$

Now replacing x with $i\theta$ in the expansion of e^x gives

$$e^{i\theta} = 1 + i\theta + \frac{(i\theta)^2}{2!} + \frac{(i\theta)^3}{3!} + \frac{(i\theta)^4}{4!} + \frac{(i\theta)^5}{5!} + \ldots$$

$$= 1 + i\theta - \frac{\theta^2}{2!} - \frac{i\theta^3}{3!} + \frac{\theta^4}{4!} + \frac{i\theta^5}{5!} + \ldots$$

$$= \cos\theta + i\sin\theta$$

Expanding a composite function

To expand a function such as $f(x) = e^{\sin x}$ we start with the expansion of e^x and replace x with $\sin x$. We can then replace $\sin x$ by its series expansion.

By terminating the series we can find a polynomial that is an approximation for the function.

Example

Find a quadratic function that is an approximation for $e^{\sin x}$.

$$e^{\sin x} = 1 + \sin x + \frac{(\sin x)^2}{2!} + \frac{(\sin x)^3}{3!} + \ldots$$

Now $\sin x = x - \frac{x^3}{3!} + \ldots$ so we can replace $\sin x$ with its expansion.

$$\therefore \quad e^{\sin x} = 1 + \left(x - \frac{x^3}{3!} + \ldots\right) + \frac{\left(x - \frac{x^3}{3!} + \ldots\right)^2}{2!} + \frac{\left(x - \frac{x^3}{3!} + \ldots\right)^3}{3!} + \ldots$$

To find a quadratic function we can ignore all terms containing powers of x greater than 2.

Now $\left(x - \frac{x^3}{3!} + \ldots\right)^2 = x^2 - \frac{2x^4}{3!} + $ higher powers of x

(and we ignore the x^4 term).

And $\left(x - \frac{x^3}{3!} + \ldots\right)^3 = x^3 + $ higher powers of x

so we can ignore this term and further terms.

$$\therefore \quad e^{\sin x} = 1 + x + \frac{x^2}{2!} + \dots$$

$$\Rightarrow \quad e^{\sin x} \simeq 1 + x + \frac{x^2}{2}$$

Using a series expansion of a function to find an approximate value of a function

By expanding a function as a Maclaurin series of ascending powers of x, we are expressing the function as an infinite polynomial. We can use the first few terms of the polynomial to find an approximate value for the function. By adding more terms we can improve on the approximation to give a value to as great a degree of accuracy as we choose, *provided* that the series converges for the value of x we use.

For example, we can find an approximate value for $\cos \frac{\pi}{4}$ by using the Maclaurin expansion of $\cos x$ which converges for all values of x.

Using the first three terms of the series, i.e. $\cos x = 1 - \frac{x^2}{2!} + \frac{x^4}{4!} - \dots$

gives $\quad \cos \frac{\pi}{4} \simeq 1 - \dfrac{\left(\frac{\pi}{4}\right)^2}{2!} + \dfrac{\left(\frac{\pi}{4}\right)^4}{4!} \dots$

$$= 0.707429\dots$$

The calculator gives $\cos \frac{\pi}{4} = 0.707106\dots$

so the approximation is correct to 3 decimal places.

Adding more terms will improve the approximation.

Adding the next term in the series, i.e. $-\dfrac{\left(\frac{\pi}{4}\right)^6}{6!}$ gives

$$\cos \frac{\pi}{4} = 1 - \dfrac{\left(\frac{\pi}{4}\right)^2}{2!} + \dfrac{\left(\frac{\pi}{4}\right)^4}{4!} - \dfrac{\left(\frac{\pi}{4}\right)^6}{6!}$$

$$= 0.707102\dots$$

and this agrees with the calculator value to 5 decimal places.

> ### Did you know?
>
> The summation of infinite series goes back to the Ancient Greeks. Archimedes used the summation of an infinite series to find the area under an arc of a parabola. He also used a series to find a fairly accurate value for π.

Exercise 2.7

1   Expand $\ln(1 + 2x^2)$ as a series of ascending powers of x as far as and including the term in x^4. Give the range of values of x for which the expansion is valid.

2   Use the first four terms of a Maclaurin series to find approximate values for:
   (a) e^2
   (b) $\ln 1.1$ (i.e. $1 + 0.1$)
   (c) $\sin \frac{\pi}{3}$

3   Write down the first five terms in the Maclaurin series expansion of e^x.
   (a) By substituting 1 for x, find an approximate value for e.
   (b) Find the value of the sixth term of the expansion when $x = 1$ and hence estimate the accuracy of your approximation.

Taylor's theorem

We have seen that we cannot expand $\ln x$ using Maclaurin's theorem.
This problem and others where the Maclaurin series does not give a valid
expansion can sometimes be overcome by using a Taylor series which
gives an expansion of $f(x)$ in ascending powers of $(x - a)$,

i.e. $f(x) = a_0 + a_1(x - a) + a_2(x - a)^2 + a_3(x - a)^3 + a_4(x - a)^4 + \dots$

The values of $a_0, a_1, a_2, \dots$ can be found using a method similar to the
one we used to find the Maclaurin series giving

$$f(x) = f(a) + f'(a)(x - a) + \frac{f''(a)(x - a)^2}{2!} + \frac{f'''(a)(x - a)^3}{3!} + \dots + \frac{f^r(a)(x - a)^r}{r!} + \dots$$

$$= \sum_{n=0}^{\infty} \frac{f^n(a)}{n!} (x - a)^n$$

This is called *Taylor's theorem* and you need to learn it.

You can assume that this series converges for values of x close to a for
any expansion you are asked to find.

Example

Find the first four terms in the Taylor expansion of $\ln (x)$ about a.

Using $f(x) = f(a) + f'(a)(x - a) + \dfrac{f''(a)(x - a)^2}{2!} + \dfrac{f'''(a)(x - a)^3}{3!} + \dots$

gives

$$f(x) = \ln (x) \qquad \text{so} \qquad f(a) = \ln a$$

$$f'(x) = \frac{1}{x} \qquad \text{so} \qquad f'(a) = \frac{1}{a}$$

$$f''(x) = -\frac{1}{x^2} \qquad \text{so} \qquad f''(a) = -\frac{1}{a^2}$$

$$f'''(x) = \frac{2}{x^3} \qquad \text{so} \qquad f'''(a) = -\frac{2}{a^3}$$

$$\therefore \quad \ln x = \ln a + \frac{x - a}{a} - \frac{(x - a)^2}{2a^2} + \frac{(x - a)^3}{3a^3} + \dots$$

Exercise 2.8a

Find the first three terms in the Taylor expansion about a of

1 $\tan x$ **2** $\sin x$ **3** $e^x \cos x$

Using Taylor series to find approximations

A Taylor series can sometimes be used when either the Maclaurin series is not valid or the series converges slowly.

Example

Find the first three terms of the expansion of $\sin x$ as a series of ascending powers of $\left(x - \frac{\pi}{4}\right)$.

Hence find an approximate value of $\sin 46°$ given that $1° = 0.017$ rad.

Using Taylor's theorem with $a = \frac{\pi}{4}$ gives

$$\sin x = \sin \frac{\pi}{4} + \left(\cos \frac{\pi}{4}\right)\left(x - \frac{\pi}{4}\right) + \left(-\sin \frac{\pi}{4}\right)\frac{\left(x - \frac{\pi}{4}\right)^2}{2!} + \ldots$$

$$= \frac{1}{\sqrt{2}} + \frac{x - \frac{\pi}{4}}{\sqrt{2}} - \frac{\left(x - \frac{\pi}{4}\right)^2}{2\sqrt{2}} + \ldots$$

Now $\sin 46° = \sin(45° + 1°) = \sin\left(\frac{\pi}{4} + 0.017\right)$

so when $x = 46°$, i.e. $\left(\frac{\pi}{4} + 0.017\right)$ rad, $\left(x - \frac{\pi}{4}\right) = 0.017$

$\therefore \quad \sin 46° = \frac{1}{\sqrt{2}} + \frac{0.017}{\sqrt{2}} - \frac{(0.017)^2}{2\sqrt{2}} + \ldots$

$\qquad\qquad = 0.707106\ldots + 0.0120208\ldots - 0.000102\ldots = 0.719025\ldots$

Therefore $\sin 46° \approx 0.7190$

($\sin 46° = 0.7193$ correct to 4 decimal places)

Note that we could use the Maclaurin series to find an approximate value for $\sin\left(\frac{\pi}{4} + 0.017\right)$ but the terms decrease in value more slowly (the third term of the expansion of $\sin\left(\frac{\pi}{4} + 0.017\right)$ is $0.00277\ldots$) so we would need more terms to give a reasonable approximation.

Using Taylor series to find polynomial approximations for the solution of differential equations

There are some differential equations that cannot be solved to give $y = f(x)$, but we can sometimes use Taylor's theorem to find a polynomial that is an approximation for $f(x)$ for values of x close to a given value.

To do this we need to know a pair of corresponding values of x and y for an equation involving $\frac{dy}{dx}$ and, for an equation involving $\frac{d^2y}{dx^2}$, corresponding values of x, y and $\frac{dy}{dx}$.

These are called the ***initial conditions***.

Then, stopping the series after a given number of terms, we can often approximate the solution for values of x close to a where a is the initial value of x. How good the approximation is depends on several things such as the number of terms included, how close x is to a, and so on.

Differential equations are usually given in terms of x, y, $\dfrac{dy}{dx}$, $\dfrac{d^2y}{dx^2}$, ..., so we use Taylor's theorem in the form

$$y = y_a + \left(\frac{dy}{dx}\right)_a (x - a) + \left(\frac{d^2y}{dx^2}\right)_a \frac{(x-a)^2}{2!} + \left(\frac{d^3y}{dx^3}\right)_a \frac{(x-a)^3}{3!} + \cdots$$

where y_a, $\left(\dfrac{dy}{dx}\right)_a$, ... means the value of y, $\dfrac{dy}{dx}$, ... when $x = a$, where a is the initial value of x.

We illustrate this with a simple first example.

Example

Find a Taylor series polynomial up to and including the term in x^2 to approximate the solution of $\dfrac{dy}{dx} = xy$ for values of x close to 0, given that $y = 1$ when $x = 0$

We stop the series after the term containing x^2,

i.e. $y \simeq y_a + \left(\dfrac{dy}{dx}\right)_a (x - a) + \dfrac{d^2y}{dx^2} \dfrac{(x-a)^2}{2!}$

The term involving x^2 involves $\dfrac{d^2y}{dx^2}$ so we differentiate the given differential equation to give an equation containing $\dfrac{d^2y}{dx^2}$

$$\frac{dy}{dx} = xy \quad \Rightarrow \quad \frac{d^2y}{dx^2} = \frac{dy}{dx} + y$$

a is the initial value of x, so $a = 0$,
and $y = 1$ when $x = 0$ so $y_a = 1$

$\therefore \quad \left(\dfrac{dy}{dx}\right)_a = (0)(1) = 0 \quad \Rightarrow \quad \left(\dfrac{d^2y}{dx^2}\right)_a = (0)(0) + 1 = 1$

$\therefore \quad y \simeq y_a + \left(\dfrac{dy}{dx}\right)_a (x - a) + \left(\dfrac{d^2y}{dx^2}\right)_a \dfrac{(x-a)^2}{2!}$

gives $y \simeq 1 + \dfrac{x^2}{2}$

The differential equation in the example above has an exact solution, $y = e^{\frac{1}{2}x^2}$

Try to judge the accuracy of the approximate solution by finding values of $1 + \dfrac{x^2}{2}$ and $e^{\frac{1}{2}x^2}$ when $x = 0.01, 0.1, 0.2$

Example

The displacement, s metres, of a particle at time t seconds is given by the differential equation

$$\frac{d^2s}{dt^2} + 2\frac{ds}{dt} + \sin s = 0$$

When $t = 0$, $s = 0$ and $\dfrac{ds}{dt} = 0.5$

Find a Taylor series approximation for s in ascending powers of t up to and including the term in t^3.

For the term in t^3 we need the value of $\frac{d^3s}{dt^3}$ so we differentiate the given differential equation.

$$\frac{d^2s}{dt^2} + 2\frac{ds}{dt} + \sin s = 0 \qquad [1]$$

$$\Rightarrow \quad \frac{d^3s}{dt^3} + 2\frac{d^2s}{dt^2} + (\cos s)\frac{ds}{dt} = 0 \qquad [2]$$

We use the Taylor series in the form

$$s = s_a + \left(\frac{ds}{dt}\right)_a (t - a) + \left(\frac{d^2s}{dt^2}\right)_a \frac{(t - a)^2}{2!} + \left(\frac{d^3s}{dt^3}\right)_a \frac{(t - a)^3}{3!} + \dots$$

Using the initial values we have $a = 0$

so $\quad s_a = 0$ and $\left(\frac{ds}{dt}\right)_a = 0.5$

Substituting these values in [1] gives

$$\left(\frac{d^2s}{dt^2}\right)_a + 2(0.5) + \sin 0 = 0$$

$$\Rightarrow \quad \left(\frac{d^2s}{dt^2}\right)_a = -1$$

Substituting these values in [2] gives

$$\left(\frac{d^3s}{dt^3}\right)_a + 2(-1) + (\cos 0)(0.5) = 0$$

$$\Rightarrow \quad \left(\frac{d^3s}{dt^3}\right)_a = \frac{3}{2}$$

Therefore $s \simeq 0 + (0.5)t + (-1)\frac{t^2}{2} + \left(\frac{3}{2}\right)\frac{t^3}{6}$

$$\Rightarrow \quad s \simeq \frac{t}{2} - \frac{t^2}{2} + \frac{t^3}{4}$$

Remember that this approximation is only reasonable when t is very small (i.e. close to zero).

Exercise 2.8b

1 Find a Taylor series approximation to y in ascending powers of x up to and including the term in x^3 when x is close to zero, given that

$$\frac{dy}{dx} = 2xy - e^x$$

and that $y = 1$ when $x = 0$

Hence find an approximate value of y when $x = 0.1$

2 Use a Taylor series expansion to find a cubic function that is approximately equal to y when

$$\frac{d^2y}{dx^2} = \left(\frac{dy}{dx}\right)^2$$

given $y = 1$ and $\frac{dy}{dx} = 2$ when $x = 1$

Hence find an approximate value of y when $x = 0.9$

Binomials

A ***binomial*** is an expression with two terms, for example, $2 + x$, $3x + 2y$, $s^2 - 5t$

In this topic we investigate how to expand powers of binomials as a series.

Pascal's triangle

We can expand, for example, $(a + b)^5$, by multiplying out the brackets, but a quicker method is to use Pascal's triangle.

First look at these expansions:

$(a + b)^1 = a + b$

$(a + b)^2 = a^2 + 2ab + b^2$

$(a + b)^3 = a^3 + 3a^2b + 3ab^2 + b^3$

$(a + b)^4 = a^4 + 4a^3b + 6a^2b^2 + 4ab^3 + b^4$

Notice that the powers of a and b form a pattern.

From the expansion of $(a + b)^4$ you can see that the first term is a^4 and then the power of a decreases by 1 in each succeeding term while the power of b increases by 1. In all the terms, the sum of the powers of a and b is 4. There is a similar pattern in the other expansions.

Now look at just the coefficients of the terms. Writing these in a triangular array gives:

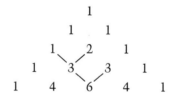

This array is called ***Pascal's triangle*** and it also has a pattern:

Each row starts and ends with 1 and each other number is the sum of the two numbers in the row above it, as shown. Also, the numbers in each row are symmetric about the middle of the row.

You can now write down as many rows as you need.

For example, to expand $(a + b)^6$, go as far as row 6:

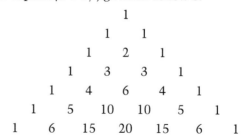

Using what we know about the pattern of the powers and using row six of the array gives

$$(a + b)^6 = a^6 + 6a^5b + 15a^4b^2 + 20a^3b^3 + 15a^2b^4 + 6ab^5 + b^6$$

The binomial theorem for $x \in \mathbb{N}$

We can use Pascal's triangle to expand $(a + b)^n$ for any $n \in \mathbb{N}$, but this will clearly be a time-consuming activity for values of n greater than 5. However, we can use Maclaurin's expansion of $(1 + x)^n$ to get a general form for the expansion of $(a + b)^n$ for any $n \in \mathbb{N}$

Using $f(x) = (1 + x)^n$,

$$f'(x) = n(x + 1)^{n-1},$$

$$f''(x) = n(n - 1)(x + 1)^{n-2},$$

$$f'''(x) = n(n - 1)(n - 2)(x + 1)^{n-3}, \dots,$$

$$f^r(x) = n(n - 1)\dots(n - r + 1)(x + 1)^r, \dots,$$

$$f^n(x) = n(n - 1)\dots(n - (n - 1))(x + 1)^0 = n(n - 1)\dots(1) = n!$$

so all further derivatives of $f(x)$ are zero and the series terminates.

$$\therefore \quad f(0) = 1, \ f'(0) = n, \ f''(0) = n(n - 1), \ f'''(0) = n(n - 1)(n - 2), \dots,$$

$$f^r(0) = n(n - 1)\dots(n - r + 1), \dots, f^n(0) = n!$$

$$\Rightarrow \quad (1 + x)^n = 1 + nx + \frac{n(n - 1)}{2!}x^2 + \frac{n(n - 1)(n - 2)}{3!}x^3 + \dots$$

$$+ \frac{n(n - 1)\dots(n - r + 1)}{r!}x^r + \dots + x^n$$

$\left(\text{Note that } \dfrac{n!}{n!} = 1\right)$

This expansion can be adapted to give the expansion of $(a + b)^n$, but before we do that, we will introduce a simpler notation for the coefficients of x, x^2, ... (These coefficients are called the **binomial coefficients**.)

The ^nC_r notation

The coefficient of x^3 in the expansion above is $\dfrac{n(n - 1)(n - 2)}{3!}$, which we can write using only factorials as $\dfrac{n!}{(n - 3)!3!}$

Similarly, $\dfrac{n(n - 1)\dots(n - r + 1)}{r!}$ can be written as $\dfrac{n!}{(n - r)!r!}$, which we denote by ^nC_r, i.e.

$$^nC_r = \frac{n!}{(n - r)!r!}$$

Therefore $\quad ^4C_2 = \dfrac{4!}{(4 - 2)!2!} = \dfrac{4 \times 3 \times 2 \times 1}{2 \times 2} = 6$

and $\quad ^8C_3 = \dfrac{8!}{(8 - 3)!3!} = \dfrac{8!}{5! \times 3!} = \dfrac{8 \times 7 \times 6}{6} = 56$

Now ^nC_r is the coefficient of x^r in the expansion of $(1 + x)^n$ so ^nC_n is the coefficient of x^n, which we know is 1, but $^nC_n = \dfrac{n!}{(n - n)!n!} = \dfrac{n!}{0!n!}$.

To make this equal to 1, we define 0! as 1

$$0! = 1$$

Example

Show that $^nC_{n-r} = {^nC_r}$

$$^nC_{n-r} = \frac{n!}{(n - (n - r))!(n - r)!} = \frac{n!}{r!(n - r)!} = \frac{n!}{(n - r)!r!} = {^nC_r}$$

Example

Find a relationship between n and r given that $^nC_r = {^{n-1}C_{r-1}}$

$$^nC_r = \frac{n!}{(n - r)!r!} \quad \text{and}$$

$$^{n-1}C_{r-1} = \frac{(n - 1)!}{(n - 1 - r + 1)!(r - 1)!} = \frac{(n - 1)!}{(n - r)!(r - 1)!}$$

$$\therefore \quad \frac{n!}{(n - r)!r!} = \frac{(n - 1)!}{(n - r)!(r - 1)!}$$

Now $n! = n(n - 1)!$ and $r! = r(r - 1)!$

Cancelling gives i.e. $\dfrac{n(n - 1)!}{(n - r)!\, r(r - 1)!} = \dfrac{(n - 1)!}{(n - r)!\, (r - 1)!}$

$$\frac{n}{r} = 1$$

$$\Rightarrow \quad n = r$$

Exercise 2.9a

1 Find the value of n when $^nC_8 = {^{n-1}C_7}$

2 Find the value of n when $5(^nC_3) = 4(^{n+1}C_3)$

3 Find a relationship between n and r given that $^{n+1}C_r = {^nC_{r+1}}$

The expansion of $(a + b)^n$ for $n \in \mathbb{N}$

We can now write the expansion of $(1 + x)^n$ as

$$^nC_0 + {^nC_1}x + {^nC_2}x^2 + \ldots + {^nC_r}x^r + \ldots + {^nC_n}x^n$$

Then $(a + b)^n = a^n\left(1 + \dfrac{b}{a}\right)^n$ and replacing x by $\dfrac{b}{a}$ in the expansion above gives

$$(a + b)^n = a^n\left({^nC_0} + {^nC_1}\left(\frac{b}{a}\right) + {^nC_2}\left(\frac{b}{a}\right)^2 + \ldots + {^nC_r}\left(\frac{b}{a}\right)^r + \ldots + {^nC_n}\left(\frac{b}{a}\right)^n\right)$$

Multiplying through by a^n and noting that $^nC_0 = {}^nC_n = 1$ gives

$$(a + b)^n = a^n + {}^nC_1 a^{n-1}b + {}^nC_2 a^{n-2}b^2 + \dots + {}^nC_r a^{n-r}b^r + \dots + b^n$$

You need to learn this, but you may find it easier to remember in the form

$$(a + b)^n = a^n + na^{n-1}b + \frac{n(n-1)}{2!} a^{n-2}b^2 + \frac{n(n-1)(n-2)}{3!} a^{n-3}b^3$$

$$+ \dots + \frac{n(n-1)\dots(n-r+1)}{r!} a^{n-r}b^r + \dots + nab^{n-1} + b^n$$

Either of these forms confirms the observations we made from Pascal's triangle, i.e. the sum of the powers of a and b in each term is n and the power of a decreases by 1 in each succeeding term while the power of b increases by 1.

We have shown that $^nC_{n-r} = {}^nC_r$ so the coefficients are symmetric about the centre.

To expand $(1 + x)^{10}$ in ascending powers of x as far as the term in x^3 we replace n with 10, a with 1 and b with x to give

$$(1 + x)^{10} = 1(1)^{10} + 10(1)^9 x + \frac{10 \times 9}{2 \times 1} (1)^8 x^2 + \frac{10 \times 9 \times 8}{3 \times 2 \times 1} (1)^7 x^3 + \dots$$

$$= 1 + 10x + 45x^2 + 120x^3 + \dots$$

Knowing the properties of the expansion, we can also write down the last four terms, i.e.

$$\dots + 120x^7 + 45x^8 + 10x^9 + x^{10}$$

To expand $(1 - x)^8$ in descending powers of x as far as the term in x^6, we can either write $(1 - x)^8$ as $(-x + 1)^8$, then replace n with 8, a with $-x$ and b with 1 to give

$$(1 - x)^8 = 1(-x)^8 + 8(-x)^7(1) + \frac{8 \times 7}{2} (-x)^6 (1)^2 \dots$$

$$= x^8 - 8x^7 + 28x^6 + \dots$$

or we can expand $(1 - x)^8$ in ascending powers and use the symmetry property, i.e.

$$(1 - x)^8 = 1 - 8x + 28x^2 + \dots + 28x^6 - 8x^7 + x^8$$

then reverse to give descending powers of x.

Note that the general term in the expansion of

$$(1 - x)^8 \text{ is } {}^8C_r(-x)^r = {}^8C_r(-1)^r x^r$$

We look at further expansions using this work in Topic 2.10.

Exercise 2.9b

1 Expand $(1 - 2x)^7$ in ascending powers of x as far as the term in x^3.

2 Find the coefficient of the term in x^4 in the expansion of $(3 - x)^5$

2.10 Applications of the binomial expansion for $n \in \mathbb{N}$

Learning outcomes

- To apply the binomial expansion for $n \in \mathbb{N}$

You need to know

- The expansion of $(a + b)^n$ for $n \in \mathbb{N}$
- The meaning of compound interest
- The sum of the first n terms of a geometric progression

The expansions of $(1 + x)^n$ and $(1 - x)^n$

$$(1 + x)^n = 1 + nx + \frac{n(n - 1)}{2}x^2 + \frac{n(n - 1)(n - 2)}{3!}x^3 + \ldots + x^n$$

and

$$(1 - x)^n = 1 - nx + \frac{n(n - 1)}{2}x^2 + \frac{n(n - 1)(n - 2)}{3!}x^3 + \ldots + (-1)^n x^n$$

These are the most straightforward binomial expansions and you need to recognise the left-hand side when you see it. For example, you should recognise $1 + 3x + 3x^2 + x^3$ as the expansion of $(1 + x)^3$

Compound interest problems

Suppose $\$A$ is deposited in an account that pays interest of $\frac{r}{100}$ of $\$A$ (where r is the rate % per annum (pa)) and the interest is credited to the account each year on the anniversary of the deposit. Then, if no withdrawals are made, at the end of year 1, the amount in the account is $\$A\left(1 + \frac{r}{100}\right)$

at the end of year 2, the amount is

$$\$A\left(1 + \frac{r}{100}\right) + \frac{r}{100} \text{ of } \$A\left(1 + \frac{r}{100}\right) = \$A\left(1 + \frac{r}{100}\right)^2$$

at the end of year 3, the amount is

$$\$A\left(1 + \frac{r}{100}\right)^2 + \frac{r}{100} \text{ of } \$A\left(1 + \frac{r}{100}\right)^2 = \$A\left(1 + \frac{r}{100}\right)^3$$

By deduction, the amount at the end of year n is $\$A\left(1 + \frac{r}{100}\right)^n$

This formula is used to calculate compound interest (where the interest is added to the capital each year).

For example, if $\$10\,000$ is deposited in an account paying 2% pa compound interest, then the amount in the account after 4 years is $\$10\,000(1 + 0.02)^4 = \$10\,000(1.02)^4$

Example

(a) Rachel has a pension that each year increases by 3% of its value the previous year. Her initial pension was $\$3000$ when she retired. What was her pension at the end of the 8th year of her retirement?

(b) How much in total did Rachel receive in pension payments for the first 8 years of her retirement?

(a) Pension at the end of the 8th year = $\$3000(1.03)^8 = \$3800$ (to the nearest $\$$)

(b) Total pension paid for the first 8 years is $\$3000(1 + 1.03 + 1.03^2 + 1.03^3 + \ldots + 1.03^8)$

The expression in brackets is the sum of the first 8 terms of a GP, with $a = 1$ and $r = 1.03$

$\therefore$ total paid $= \$3000 \times \dfrac{1(1 - 1.03^8)}{1 - 1.03} = \$3000 \times \dfrac{1.03^8 - 1}{0.03} = \$26\,677$ (to the nearest $\$$)

Expansions using the binomial theorem

The examples that follow illustrate some problems involving expansions.

Example

Find the terms up to and including x^3 in the expansion of $(1 + 2x)^4\left(1 - \frac{1}{2}x\right)^6$

$$(1 + 2x)^4\left(1 - \frac{1}{2}x\right)^6 = \left(1 + 4(2x) + 6(2x)^2 + 4(2x)^3 + \ldots\right)\left(1 - 6\left(\frac{1}{2}x\right) + \frac{6 \times 5}{2!}\left(\frac{1}{2}x\right)^2 - \frac{6 \times 5 \times 4}{3!}\left(\frac{1}{2}x\right)^3 + \ldots\right)$$

There is no need to go beyond the term in x^3 in either expansion

$$= \left(1 + 8x + 24x^2 + 32x^3 + \ldots\right)\left(1 - 3x + \frac{15}{4}x^2 - \frac{5}{2}x^3 + \ldots\right)$$

$$= 1 - 3x + \frac{15}{4}x^2 - \frac{5}{2}x^3 + \ldots$$

$$+ 8x - 24x^2 + 30x^3 + \ldots$$

$$+ 24x^2 - 72x^3 + \ldots$$

$$+ 32x^3 + \ldots$$

$$1 + 5x + \frac{15}{4}x^2 - 12\frac{1}{2}x^3 + \ldots$$

☑ *Exam tip*

Be systematic when you expand brackets like this: multiply the second bracket by 1, then by 8x and so on. Then add the results.

Example

Find the term independent of x in the expansion of $\left(\frac{1}{x} - 2x^2\right)^9$

The general term in the expansion of $\left(\frac{1}{x} - 2x^2\right)^9$ is

$$^9C_r\left(\frac{1}{x}\right)^{9-r}(-2x^2)^r = {}^9C_r\,(-2)^r\left(\frac{x^{2r}}{x^{9-r}}\right)$$

This term is independent of x when $2r = 9 - r$, i.e. when $r = 3$

Therefore the term independent of x is $^9C_3(-2)^3 = \frac{9! \times (-8)}{3!6!} = -672$

Example

Use the binomial expansion of $(1 - 2x)^8$ to find the value of 0.98^8 correct to 3 decimal places.

$$(1 - 2x)^8 = 1 - 8(2x) + 28(4x^2) - 56(8x^3) + 70(16x^4) - 56(32x^5) + \ldots$$

$$0.98 = 1 - 2(0.01)$$

So substituting 0.01 for x gives

$$0.98^8 = 1 - 0.16 + 0.0112 - 0.000448 + 0.0000112 - 0.0000001792 + \ldots$$

We stop here as the first significant figure of the next term will be in the 7th or 8th decimal place so will not alter the 4th decimal place

$$= 0.851 \text{ correct to 3 d.p.}$$

Exercise 2.10

1. Find the coefficient of x^3 in the expansion of $(1 + x - x^2)^6$
   (Hint: treat it as $(1 + X)^6$ and then substitute $x - x^2$ for X.)

2. Find the real part of $(1 + 2i)^6$
   (Hint: expand $(1 + x)^6$, replacing x by 2i and only consider even powers of i.)

3. Find the coefficient of x^7 in the expansion of $\left(x^2 - \frac{1}{x}\right)^8$

Learning outcomes

- To derive the expansion of $(1 + x)^n$ when n is a fraction or a negative integer
- To apply the binomial theorem to problems

You need to know

- Maclaurin's theorem
- Factorial notation
- The meaning of a convergent series
- How to express a rational function in partial fractions

The binomial theorem for $n \in \mathbb{Q}$

Using Maclaurin's theorem to expand $(1 + x)^n$ gives

$$(1 + x)^n = 1 + nx + \frac{n(n - 1)}{2!}x^2 + \frac{n(n - 1)(n - 2)}{3!}x^3$$

$$+ \ldots + \frac{n(n - 1)\ldots(n - r + 1)}{r!}x^r + \ldots$$

Now r is a positive integer, but when n is a negative integer or a fraction, there is no value of r for which $(n - r + 1)$ is zero. In this case the series does not terminate.

The series expansion of $(1 + x)^n$ converges to $(1 + x)^n$ only when $-1 < x < 1$

Note that *the term in x^n is the $(n + 1)$th term*, not the nth term.

For example, to expand $(1 + x)^{\frac{1}{2}}$ we substitute $\frac{1}{2}$ for n giving

$$(1 + x)^{\frac{1}{2}} = 1 + \frac{1}{2}x + \frac{\frac{1}{2} \times \left(-\frac{1}{2}\right)}{2!}x^2 + \frac{\frac{1}{2} \times \left(-\frac{1}{2}\right) \times \left(-\frac{3}{2}\right)}{3!}x^3 + \ldots$$

$$= 1 + \frac{1}{2}x - \frac{1}{8}x^2 + \frac{1}{16}x^3 + \ldots \qquad \text{for } -1 < x < 1$$

Note that when n is a positive integer, the series $(1 + x)^n$ terminates and is valid for all values of x, but when n is not a positive integer, the series is infinite and converges only when $|x| < 1$.

There are other differences:

we cannot use $^n\mathrm{C}_r$ for the coefficients

and we cannot use the form of the expansion for $(a + b)^n$

To expand $(a + b)^n$ when $n \notin \mathbb{N}$, we take a outside the bracket to give

$$a^n \left(1 + \frac{b}{a}\right)^n$$

For example, to expand $\sqrt{(2 - x^2)}$, we express $\sqrt{(2 - x^2)}$ as $2^{\frac{1}{2}} \left(1 - \frac{x^2}{2}\right)^{\frac{1}{2}}$

then replacing n by $\frac{1}{2}$ and x by $\left(-\frac{x^2}{2}\right)$ we have

$$2^{\frac{1}{2}} \left(1 - \frac{x^2}{2}\right)^{\frac{1}{2}} = 2^{\frac{1}{2}} \left(1 + \left(\frac{1}{2}\right)\left(-\frac{x^2}{2}\right) + \frac{\frac{1}{2} \times \left(-\frac{1}{2}\right)}{2!}\left(-\frac{x^2}{2}\right)^2 + \ldots\right)$$

$$= \sqrt{2} - \frac{x^2\sqrt{2}}{4} - \frac{x^4\sqrt{2}}{32} - \ldots$$

This expansion is valid when $-1 < \frac{x^2}{2} < 1$

i.e.   when $0 < \frac{x^2}{2} < 1$ $\frac{x^2}{2}$ cannot be negative

$\Rightarrow \quad x^2 < 2 \quad \Rightarrow \quad -\sqrt{2} < x < \sqrt{2}$

Example

Expand $(1 + x)^{-1}$ as a series of ascending powers of x up to and including the term in x^4.

Give the term in x^n.

Using the binomial theorem,

$(1 + x)^{-1}$

$= 1 - x + \dfrac{(-1)(-2)}{2!}x^2 + \dfrac{(-1)(-2)(-3)}{3!}x^3 + \dfrac{(-1)(-2)(-3)(-4)}{4!}x^4 + \ldots$

$= 1 - x + x^2 - x^3 + x^4 + \ldots$

The pattern is now clear, i.e. the coefficients are 1 when the power of x is even and -1 when the powers of x are odd.

Therefore the term in x^n is $(-1)^n x^n$

The series expansion of $(1 - x)^{-1}$ is similar to the series in the example above, i.e.

$$(1 - x)^{-1} = 1 - (-x) + \dfrac{(-1)(-2)}{2!}(-x^2) + \dfrac{(-1)(-2)(-3)}{3!}(-x^3)$$

$$+ \dfrac{(-1)(-2)(-3)(-4)}{4!}(-x)^4 + \ldots$$

$$= 1 + x + x^2 + x^3 + x^4 + \ldots$$

The series expansions of $(1 + x)^{-1}$ and $(1 - x)^{-1}$ are worth remembering and may be quoted unless their derivation is asked for, i.e.

$(1 + x)^{-1} = 1 - x + x^2 - x^3 + x^4 + \ldots + (-1)^n x^n + \ldots \quad -1 < x < 1$

$(1 - x)^{-1} = 1 + x + x^2 + x^3 + x^4 + \ldots + x^n + \ldots \quad -1 < x < 1$

Note that both of these series are geometric, so starting with the right-hand side and finding the sum to infinity of a GP verifies these expansions.

Exercise 2.11a

1   Expand $(1 - x)^{\frac{1}{2}}$ as a series of ascending powers of x as far as the term in x^5.
   Give the range of values for which the expansion is valid.

2   (a) Expand $(1 + 3x)^{\frac{1}{2}}$ as a series of ascending powers of x as far as the term in x^5.
   (b) Find the term in x^n and give the range of values for which the expansion is valid.

3   Find the term in x^n in the binomial expansion of $(1 - 2x)^{-2}$ and give the range of values of x for which the expansion is valid.

Applications of the binomial theorem

We can apply the binomial theorem to a variety of functions if we can express them as binomials.

To use the binomial theorem to expand a function such as $f(x) = (x - 1)^{-1}$ we write it as $x^{-1}\left(1 - \frac{1}{x}\right)^{-1}$. We can then expand the function as a series of descending powers of x,

i.e. $\quad f(x) = \frac{1}{x}\left(1 + \frac{1}{x} + \left(\frac{1}{x}\right)^2 + \left(\frac{1}{x}\right)^3 + \dots\right) = \frac{1}{x} + \frac{1}{x^2} + \frac{1}{x^3} + \dots$

$\qquad\qquad = x^{-1} + x^{-2} + x^{-3} + \dots$

This series is valid for $-1 < \frac{1}{x} < 1$, i.e. for $x < -1$ or $x > 1$

Example

Find the coefficient of x^n in the expansion of $(3 - 2x)^{-2}$ in ascending powers of x and give the range of values of x for which the expansion is valid.

$(3 - 2x)^{-2} = 3^{-2}\left(1 - \frac{2x}{3}\right)^{-2}$

$\qquad = \frac{1}{9}\left(1 + (-2)\left(-\frac{2x}{3}\right) + \frac{(-2)(-3)}{2!}\left(-\frac{2x}{3}\right)^2 + \frac{(-2)(-3)(-4)}{3!}\left(-\frac{2x}{3}\right)^3 + \frac{(-2)(-3)(-4)(-5)}{4!}\left(-\frac{2x}{3}\right)^4 + \dots\right)$

$\qquad = \frac{1}{9}\left(1 + 2\left(\frac{2x}{3}\right) + 3\left(\frac{2x}{3}\right)^2 + 4\left(\frac{2x}{3}\right)^3 + 5\left(\frac{2x}{3}\right)^4 + \dots\right)$

You need to write down sufficient terms so that the pattern of the coefficients is clear.

From this we can see that the coefficient of x^n is $\frac{1}{9}(n + 1)\frac{2^n}{3^n} = \frac{2^n(n + 1)}{3^{n+2}}$

The expansion is valid for $-1 < \frac{2x}{3} < 1 \quad \Rightarrow \quad -\frac{3}{2} < x < \frac{3}{2}$

The binomial theorem can be used to expand rational functions with factors in the denominator by using partial fractions to express them as the sum or difference of simpler functions.

Example

(a) Express $f(x) = \dfrac{1}{(1 + x^2)(1 - x)}$ in partial fractions.

(b) Hence find the first four terms in the expansion of $f(x)$ as a series of ascending powers of x, stating the range of values of x for which the expansion is valid.

(c) Find the coefficient of x^r.

(a) $\dfrac{1}{(1 + x^2)(1 - x)} \equiv \dfrac{Ax + B}{1 + x^2} + \dfrac{C}{1 - x}$

$\qquad\qquad \Rightarrow \quad 1 \equiv (Ax + B)(1 - x) + C(1 + x^2)$

$\qquad \therefore \quad C = \frac{1}{2}, \ B = \frac{1}{2}$ and $A = \frac{1}{2}$

$\qquad\qquad \Rightarrow \quad \dfrac{1}{(1 + x^2)(1 - x)} \equiv \dfrac{x + 1}{2(1 + x^2)} + \dfrac{1}{2(1 - x)}$

(b) $f(x) = \frac{1}{2}(x + 1)(1 + x^2)^{-1} + \frac{1}{2}(1 - x)^{-1}$

$\qquad = \frac{1}{2}(x + 1)(1 - x^2 + x^4 - x^6 + \ldots) + \frac{1}{2}(1 + x + x^2 + x^3 + x^4 \ldots)$

$\qquad = \frac{1}{2}(x - x^3 + x^5 - x^7 + \ldots) + \frac{1}{2}(1 - x^2 + x^4 - x^6 + \ldots)$

$\qquad\quad + \frac{1}{2}(1 + x + x^2 + x^3 + x^4 + \ldots)$ The terms in x^2 and x^3

$\qquad = \frac{1}{2}(1 + x - x^2 - x^3 + x^4 + x^5 - \ldots)$ cancel so we have to add another term to the

$\qquad\quad + \frac{1}{2}(1 + x + x^2 + x^3 + x^4 + x^5 + \ldots)$ expansion of $(1 - x)^{-1}$

$\qquad = 1 + x + x^4 + x^5 + \ldots$

The series is valid for $-1 < x < 1$

(c) The coefficient of x^r is 1.

The binomial theorem can also be used to find approximate values for some irrational numbers.

Example

(a) Expand $(1 - x)^{\frac{1}{2}}$ as far as the term in x^3.

(b) Substitute 0.02 for x in $(1 - x)^{\frac{1}{2}}$ and its expansion.
Hence find an approximate value for $\sqrt{2}$ and state the degree of accuracy of your answer.

(a) $(1 - x)^{\frac{1}{2}} = 1 - \frac{1}{2}x + \dfrac{\left(\frac{1}{2}\right)\left(-\frac{1}{2}\right)(-x^2)}{2!} + \dfrac{\left(\frac{1}{2}\right)\left(-\frac{1}{2}\right)\left(-\frac{3}{2}\right)(-x)^3}{3!} + \ldots$

$\qquad\quad = 1 - \frac{1}{2}x - \frac{1}{8}x^2 - \frac{1}{16}x^3 - \ldots$

(b) Substituting 0.02 for x gives

$\qquad (0.98)^{\frac{1}{2}} = 1 - 0.01 - 0.000\,05 - 0.000\,000\,5 - \ldots$

This expansion is valid because $x = 0.02$ is within the range $-1 < x < 1$

$\Rightarrow \quad \sqrt{\dfrac{98}{100}} = 0.989\,949\,5\ldots$ This is correct to 7 d.p. as the next term is 5×10^{-9}

$\Rightarrow \quad \dfrac{7}{10}\sqrt{2} = 0.989\,949\,5\ldots \quad \Rightarrow \quad \sqrt{2} = 1.41421$ correct to 5 d.p.

Exercise 2.11b

1 Expand $(x - 2)^{\frac{1}{2}}$ as a series of descending powers of x as far as and including the fourth term. Give the range of values of x for which the expansion is valid.

2 Express $\dfrac{1}{(1 + x)(1 - 3x)}$ in partial fractions. Hence expand

$\dfrac{1}{(1 + x)(1 - 3x)}$ as a series of ascending powers of x as far as and

including the term in x^4. Give the range of values of x for which the expansion is valid.

3 Use the expansion of $\dfrac{1}{\sqrt{1 - x}}$ with $x = 0.1$ to find the value of $\sqrt{10}$ correct to 4 d.p.

The intermediate value theorem

Consider a function $f(x)$ that is continuous between $x = a$ and $x = b$

The diagram shows that if $f(c)$ is a value of $f(x)$ between $f(a)$ and $f(b)$, then c lies between a and b.

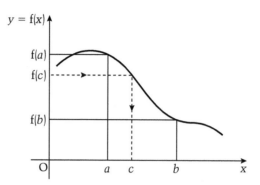

There may be more than one value of x between $x = a$ and $x = b$, as the diagram below shows.

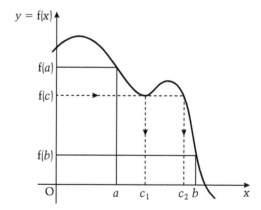

However, if $f(x)$ is not continuous between a and b then there may not be a value of x between $x = a$ and $x = b$

This is illustrated in the next diagram.

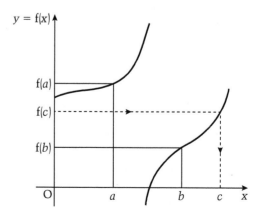

The intermediate theorem states that:

> **Provided that f(x) is continuous between $x = a$ and $x = b$, there must be at least one value of x between $x = a$ and $x = b$ for which f(x) has a given value between f(a) and f(b).**

Locating a root of an equation

It is not possible to find the exact roots of some equations.

However, we can sometimes use the intermediate theorem to locate a root in an interval.

If the equation $f(x) = 0$ has a root between $x = a$ and $x = b$, and if $f(x)$ is continuous in this interval, then the curve $y = f(x)$ crosses the x-axis between $x = a$ and $x = b$

The intermediate value theorem tells us that $f(x) = 0$ is between $f(a)$ and $f(b)$.

Therefore $f(x)$ changes sign between $x = a$ and $x = b$, i.e.

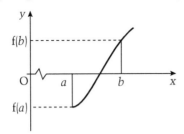

> **If f(x) is continuous between $x = a$ and $x = b$ and if one root of the equation f(x) = 0 lies between $x = a$ and $x = b$ then f(a) and f(b) are opposite in sign, i.e. f(a) × f(b) < 0**

The first step is to roughly locate the roots of an equation using a sketch where possible.

For example, the equation $e^x - 2x - 2 = 0$ has roots where the graphs of $y = e^x$ and $y = 2x + 2$ intersect.

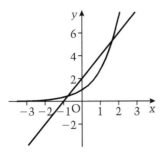

From the sketch, we can see that there appears to be a root between $x = 1$ and $x = 2$

We can test this by using $f(x) = e^x - 2x - 2$ (which is continuous) and finding $f(1)$ and $f(2)$.

$$f(1) = e - 4 < 0$$

and $\quad f(2) = e^2 - 6 > 0 \qquad\qquad e^2 = 7.3...$

As $f(1)$ and $f(2)$ are opposite in sign, i.e. $f(1) \times f(2) < 0$, there is a root between 1 and 2.

Example

Use a sketch to show that the equation $x^3 - 2x^2 + x + 1 = 0$ has only one real root.

Find two consecutive integers between which this root lies.

The roots of $x^3 - 2x^2 + x + 1 = 0$ are the values of x where the graph of $y = x^3 - 2x^2 + x + 1$ intersects the x-axis.

The curve is a cubic which crosses the y-axis where $y = 1$, and $y \to \infty$ as $x \to \infty$

To locate the curve we will find the turning points:

$$\frac{dy}{dx} = 3x^2 - 4x + 1$$

$$\Rightarrow \quad 3x^2 - 4x + 1 = 0$$

$$\Rightarrow \quad (3x - 1)(x - 1) = 0$$

$$\Rightarrow \quad x = \tfrac{1}{3} \text{ and } x = 1$$

When $x = \tfrac{1}{3}$, $y = 1\tfrac{4}{27}$ and when $x = 1$, $y = 1$

The curve crosses the x-axis once so there is only one real root.

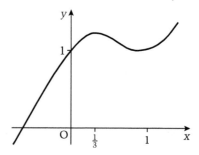

Alternatively, the roots of $x^3 - 2x^2 + x + 1 = 0$ are where $x^3 = 2x^2 - x - 1$

A sketch of the curves $y = x^3$ and $y = 2x^2 - x - 1 \ (= (2x + 1)(x - 1))$ also shows that there is only one real root.

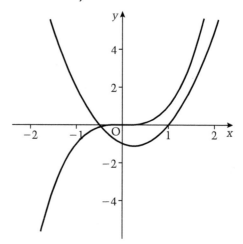

From either sketch it appears that the root is between -1 and 0.

Using $f(x) = x^3 - 2x^2 + x + 1$,

$f(-1) = -3$ and $f(0) = 1$

$f(-1)$ and $f(0)$ are opposite in sign, therefore the root lies between -1 and 0.

(It is likely that this root is nearer 0 than -1, as $f(0)$ is nearer zero than is $f(-1)$. We can test this by finding the sign of $f\left(-\frac{1}{2}\right)$.

If $f\left(-\frac{1}{2}\right) < 0$, the root lies between $-\frac{1}{2}$ and 0.)

Exercise 2.12

1 Draw a sketch to show that the equation $\ln x = \frac{1}{x}$ has one real root.

   Hence find two consecutive integers between which the root of the equation $x \ln x - 1 = 0$ lies.

2 Show, using a sketch or otherwise, that the equation $e^{-x} = x^2 - 1$ has only one root.

   Find two consecutive integers between which this root of the equation lies.

3 The diagram shows a sketch of the curve
   $$y = \tan^{-1} x - \ln(1 + x^2)$$

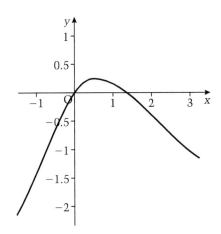

   (a) Verify that zero is one root of the equation
      $$\tan^{-1} x - \ln(1 + x^2) = 0$$

   (b) Use the intermediate value theorem to show that another root of the equation lies between 1 and 1.5

Numerical methods for solving equations

Numerical methods use repeated applications of a method to successively improve on an approximation for a root of an equation.

Interval bisection method

In the last topic, we showed how to locate a root between successive integers. The **interval bisection method** refines this approach to give the value of the root to any degree of accuracy.

Consider again the equation $e^x - 2x - 2 = 0$

We have shown in Topic 2.12 that this equation has a root between 1 and 2 and that $f(1) < 0$ and $f(2) > 0$

So if this root is α we know that $1 < \alpha < 2$

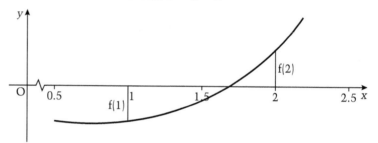

We then bisect the interval to give $x = 1.5$ and find the sign of $f(1.5)$:
$f(1.5) = e^{1.5} - 5 = -0.5...$ Therefore $f(1.5) < 0$ and $f(2) > 0$ so
$1.5 < \alpha < 2$

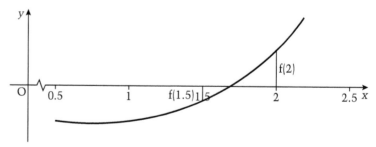

We then bisect the interval again to give $x = 1.75$ and find the sign of $f(1.75)$:
$f(1.75) = e^{1.75} - 5.5 = 0.25... > 0$

Therefore $f(1.5) < 0$ and $f(1.75) > 0$ so $1.5 < \alpha < 1.75$

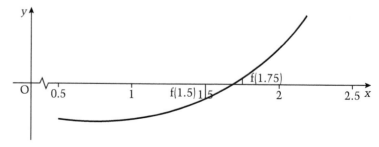

Bisecting the interval 1.5 to 1.75 gives $x = 1.625$ and
$f(1.625) = -0.17... < 0$

Therefore $f(1.625) < 0$ and $f(1.75) > 0$ so $1.625 < \alpha < 1.75$

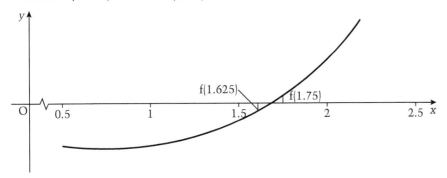

We are narrowing down the interval in which the root lies, but we still do not have a value correct to even 1 decimal place, so we continue.

There is no need to draw diagrams, we just need to keep track of the sign of $f(x)$.

Bisecting the interval again gives $x = 1.6875$

$f(1.6875) = 0.03... > 0$ and we know that $f(1.625) < 0$

Therefore $1.625 < \alpha < 1.6875$

Bisecting the interval again gives $x = 1.65625$

$f(1.65625) = -0.072... < 0$ and $f(1.6875) > 0$

Therefore $1.65625 < \alpha < 1.6875$

The last interval is less than 0.05, so we can now say that the root of the equation is 1.7 correct to 1 decimal place.

To get an answer correct to 2 decimal places, we need the interval to be less than 0.005

This method is an example of an *iterative method*. An iterative method for finding a root of an equation starts with a first approximation and then uses that to feed into the next step to give a better approximation. This is then repeated until the desired degree of accuracy is obtained. Each step is called an iteration.

The interval bisection method is slow to converge (i.e. to get close to the value of the root).

In the example above it took five iterations to get an answer correct to 1 decimal place.

However, it does have the advantage that the method will only fail if the conditions for the intermediate value theorem are not met, i.e. if the function is not continuous or there is more than one root in the initial interval.

In the next topic we look at an iteration method that improves on the interval bisection method.

Exercise 2.13

1 (a) Find the stationary points on the curve $y = x^3 - 3x + 4$ and hence sketch the curve.

   (b) Use your sketch to find consecutive intervals in which the root of the equation $x^3 - 3x + 4$ lies.

   (c) Use the interval bisection method to find this root correct to 1 decimal place.

Linear interpolation

Linear interpolation is similar to the interval bisection method but uses proportion to find the next value in the interval rather than taking the mid-point.

Consider an equation $f(x) = 0$ which has a root α which we know lies between $x = a$ and $x = b$

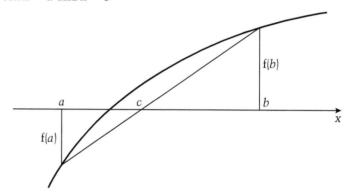

The line joining the points on the curve $y = f(x)$ where $x = a$ and $x = b$ cuts the x-axis at c. Assuming that $f(a) < 0$ and $f(b) > 0$, the diagram shows that the interval between $x = a$ and $x = c$ is likely to be smaller than the interval from $x = a$ to the interval bisection point. Therefore this method is likely to converge more quickly than the interval bisection method.

The line joining the points on the curve $y = f(x)$ where $x = a$ and $x = b$ forms a pair of similar triangles. Therefore the point $x = c_1$ divides the line between $x = a$ and $x = b$ in the ratio

$$|f(a)| : |f(b)|, \quad \text{i.e.} \quad \frac{c_1 - a}{b - c_1} = \frac{|f(a)|}{|f(b)|}$$

$$\Rightarrow \quad c_1 = \frac{a|f(b)| + b|f(a)|}{|f(a)| + |f(b)|}$$

where c_1 is the first approximation for α.

You need to learn this.

Consider again the equation $e^x - 2x - 2 = 0$

We have shown in Topic 2.12 that this equation has a root between 1 and 2 and that $f(1) < 0$ and $f(2) > 0$

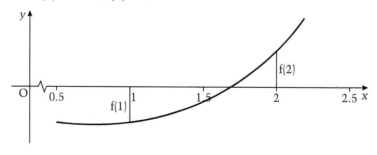

Working with the first four decimal places throughout gives

$f(1) = -1.2817...$ and $f(2) = 1.3890...$

Therefore
$$c_1 = \frac{(1)(1.3890) + (2)(1.2817)}{1.2817 + 1.3890} = 1.4799$$

This is the 1st approximation for α

$f(c_1) = -0.5672... < 0$ so α is in the interval 1.4799 to 2

Repeating the process:
$$c_2 = \frac{(1.4799)(1.3890) + (2)(0.5672)}{0.5672 + 1.3890} = 1.6307$$

2nd approximation

$f(c_2) = -0.1539... < 0$ so α is in the interval 1.6307 to 2

Repeating again:
$$c_3 = \frac{(1.6307)(1.3890) + (2)(0.1539)}{0.1539 + 1.3890} = 1.6675$$

3rd approximation

$f(c_3) = -0.0360... < 0$ so α is in the interval 1.6675 to 2

And again:
$$c_4 = \frac{(1.6675)(1.3890) + (2)(0.0360)}{0.0360 + 1.3890} = 1.6759$$

4th approximation

$f(1.6759) = -0.0081...$ and this is small enough to be worth checking to see if the 4th approximation is correct to 2 decimal places:

$f(1.675) = -0.011...$ and $f(1.685) = 0.022...$ so $1.675 < \alpha < 1.685$

Therefore $\alpha = 1.68$ correct to 2 decimal places.

We have found the value of α correct to 2 decimal places in four iterations. This compares with five iterations to give a value correct to 1 decimal place using interval bisection (Topic 2.13). Therefore the convergence rate is quicker.

The rate of convergence of linear interpolation depends on the shape of the curve in the initial interval.

If the gradient changes a great deal and if c_1 is not very close to α, the rate of convergence is slow.

If the gradient does not change much and if c_1 is close to α, the rate of convergence is fast.

As with interval bisection, this method fails if the function is not continuous or has more than one root in the initial interval.

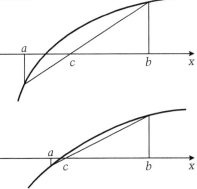

Exercise 2.14

1 (a) Show that the equation $\ln x - x + 2 = 0$ has a root between $x = 3$ and $x = 3.5$

   (b) Use linear interpolation twice to get an approximate value for this root.

   (c) Show that the approximation is correct to 3 decimal places.

2.15 Newton–Raphson method

Learning outcomes

- To use the Newton–Raphson method to find an approximate value for the root of an equation to any degree specified

- To give a geometric interpretation of the method

You need to know

- How to locate an interval in which a root of an equation lies

- How to find the equation of a tangent to a curve at a given point

Did you know?

The method was first published by Sir Isaac Newton and it is often called simply Newton's method. However, it was simplified by Joseph Raphson a few years later. Neither of these early methods used calculus – this was first introduced by Thomas Simpson. The version we use today was published nearly a century later by the French mathematician Joseph Lagrange.

Newton–Raphson method

The **Newton–Raphson method** uses a linear approximation for a function.

If the equation $f(x) = 0$ has a root α then the curve $y = f(x)$ cuts the x-axis where $x = \alpha$

If c_1 is an approximate value of α, then the tangent to the curve at the point A where $x = c_1$ cuts the x-axis at a point where $x = c_2$

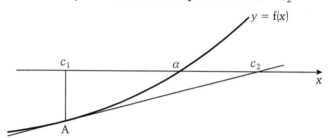

In most cases, c_2 will be closer to α than is c_1. Therefore c_2 is a better approximation to α.

The coordinates of A are $(c_1, f(c_1))$ and the gradient of the curve at A is $f'(c_1)$.

Therefore the equation of the tangent is $y - f(c_1) = f'(c_1)(x - c_1)$

This tangent cuts the x-axis where $y = 0 \Rightarrow x = c_1 - \dfrac{f(c_1)}{f'(c_1)}$

Therefore if c_1 is an approximation for a root of an equation $f(x) = 0$ then $c_2 = c_1 - \dfrac{f(c_1)}{f'(c_1)}$ is a better approximation.

You also need to learn this.

Using this method to find the root of $e^x - 2x - 2 = 0$ and using $c_1 = 2$ as the first approximation, we have $f'(x) = e^x - 2$

Therefore $\quad c_2 = 2 - \dfrac{e^2 - 6}{e^2 - 2} = 1.74224...$

and $\quad c_3 = 1.74224... - \dfrac{e^{1.74224} - 5.48449...}{e^{1.74224} - 2} = 1.68142...$

and $\quad c_4 = 1.68142... - \dfrac{e^{1.68142} - 5.36284...}{e^{1.68142} - 2} = 1.67835...$

so α is probably equal to 1.68 correct to 2 decimal places.

We have already tested this in Topic 2.14, so we know that $\alpha = 1.68$ correct to 2 decimal places.

If we do another iteration, we get

$$c_5 = 1.67835... - \dfrac{e^{1.67835} - 5.3567...}{e^{1.67835} - 2} = 1.67834...$$

so we can see that α is likely to be 1.6783 to 4 decimal places.

We can check whether $\alpha = 1.6783$ is correct to 4 decimal places:
$f(1.67825) = -0.0003..., f(1.67835) = 0.00001...$ Therefore
$1.67825 < \alpha < 1.67835$ so $\alpha = 1.6783$ correct to 4 decimal places.

The Newton−Raphson method is the best method considered so far for finding a root of an equation because when it works it converges rapidly, as the example above shows.

However, there are factors that cause the method to fail:

■ the first approximation, c_1, is too far from α

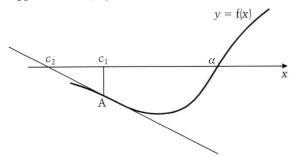

■ the gradient of the curve at the point where $x = c_1$ is too small

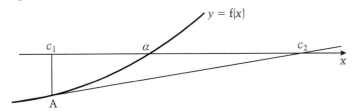

■ the gradient of the curve increases rapidly

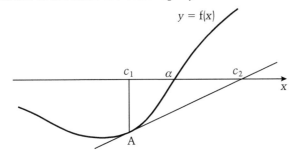

Exercise 2.15

1 **(a)** Use a sketch to show that the equation $x^2 = \ln(x + 2)$ has two roots.
  (b) Use the Newton−Raphson method three times to find an approximate value of the larger root.
  (c) State, with reasons, the accuracy of your approximation.

Iteration

As we have seen with linear interpolation and Newton–Raphson, iteration produces a sequence of values by using a formula (called an iteration formula) of the form

$$x_{n+1} = f(x_n)$$

Taking x_1 as the first value, then
$$x_2 = f(x_1)$$
$$x_3 = f(x_2)$$
$$x_4 = f(x_3)$$
$$x_5 = f(x_4) \quad \text{and so on.}$$

For example, when $x_{n+1} = (x_n + 1)^{\frac{1}{2}}$ and $x_1 = 2$

$$x_2 = (2 + 1)^{\frac{1}{2}} = 1.732\ldots$$
$$x_3 = (1.732\ldots + 1)^{\frac{1}{2}} = 1.652\ldots$$
$$x_4 = (1.652\ldots + 1)^{\frac{1}{2}} = 1.628\ldots$$
$$x_5 = (1.628\ldots + 1)^{\frac{1}{2}} = 1.621\ldots \quad \text{and so on.}$$

This is the same as a recurrence formula used to generate a sequence and we now look at the convergence of such a sequence in the context of finding a root of an equation.

The sequence of values generated above converge to a value α, because as n increases, x_n gets closer and closer to x_{n+1}, i.e. $x_n \to \alpha$. This value, α, is when $x_n = x_{n+1}$, i.e. when $\alpha = (\alpha + 1)^{\frac{1}{2}}$. Therefore α is a root of the equation $\alpha = (\alpha + 1)^{\frac{1}{2}}$

Not all iterations give values that converge.

For example, using the iteration formula, $x_{n+1} = \sqrt{e^{x_n} + 2}$, and taking $x_1 = 2$ gives

$$x_2 = \sqrt{e^2 + 2} = 3.064\ldots$$
$$x_3 = \sqrt{e^{3.064\ldots} + 2} = 4.839\ldots$$
$$x_4 = \sqrt{e^{4.839\ldots} + 2} = 11.32\ldots$$

This sequence of values diverges because the values are increasing (rapidly in this case).

Using an iteration formula to find a root

We have seen that we can use an iteration formula to find a good approximation to a root, α, of an equation $f(x) = 0$

When $f(x) = 0$ can be written in the form $x = g(x)$ we can use this to make the iteration formula

$$x_{n+1} = g(x_n)$$

The roots of the equation $x = g(x)$ are the values of x at the points of intersection of the line

$y = x$ and the curve $y = g(x)$

The diagram shows how this iteration works.

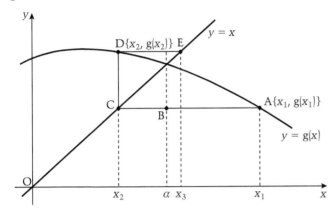

Using x_1 as the first approximation to the root α, then in the diagram

A is the point on $y = g(x)$ where $x = x_1$ so $y = g(x_1)$

B is the point where $x = \alpha$ and $y = g(x_1)$

C is the point on the line $y = x$ where $x = x_2$ and $y = g(x_1)$

Now x_2 will be closer to α than is x_1 provided that, near the root, the curve is less steep than the gradient of the line $y = x$, i.e. provided that $|g'(x)| < 1$

Therefore x_2 is a better approximation to α than is x_1, provided that $|g'(x)| < 1$

Now C is on the line $y = x$, therefore $x_2 = g(x_1)$

We can repeat this process to get $x_3, x_4, \ldots$.

The rate of convergence of this sequence depends on the value of $g'(x)$ near the root.

The smaller $|g'(x)|$ is, the more rapid is the rate of convergence.

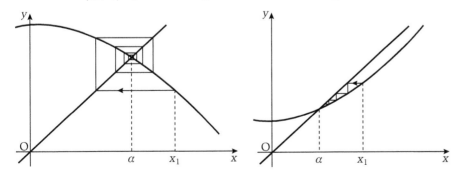

The sequence diverges (i.e. fails to find a root) if $|g'(x)| > 1$

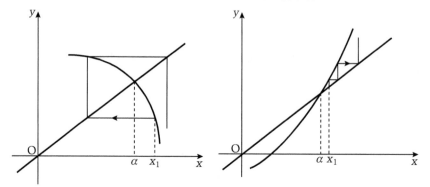

We will use this method to try and find the roots of the equation
$e^{x+1} - x - 3 = 0$

The graph of $y = e^{x+1} - x - 3$ shows that the equation has two roots, one near -3 and the other near 0.

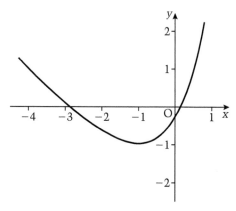

Rearranging the equation as $x = e^{x+1} - 3$ and changing this to the iteration formula gives

$x_{n+1} = e^{x_n + 1} - 3$

Taking $x_1 = -3$ gives

$$x_2 = e^{-2} - 3 = -2.8646...$$
$$x_3 = e^{-1.8646...} - 3 = -2.8450...$$
$$x_4 = e^{-1.8450...} - 3 = -2.8419...$$
$$x_5 = e^{-1.8419...} - 3 = -2.8414...$$

so this iteration is converging.

Using $f(x) = e^{x+1} - x - 3$,
$f(-2.8415) = 7.9 \times 10^{-5} > 0$ and $f(-2.8405) = -7.6 \times 10^{-4} < 0$,
therefore $\alpha = -2.841$ correct to 4 significant figures.

Now taking $x_1 = 0$ as the first approximation to the other root gives

$$x_2 = e^1 - 3 = -0.2817...$$
$$x_3 = e^{1 - 0.2817...} - 3 = -0.9490...$$
$$x_4 = e^{1 - 0.9490...} - 3 = -1.9477...$$

This sequence is diverging so it fails to find the root near zero.

We could predict that this will happen by looking at the gradient function of $e^{x+1} - 3$:

$$\frac{d}{dx}(e^{x+1} - 3) = e^{x+1} \text{ and } e^{x+1} > 1 \text{ for } x > -1,$$

i.e. $|g'(x)| > 1$ for values of x near $x = 0$

Example

(a) Show that the equation $\cos x - x = 0$ has a root between 0 and $\frac{\pi}{3}$

(b) Taking 0.75 as a first approximation to this root, use the iteration $x_{n+1} = \cos x_n$ three times to find an approximation to this root.

(c) Hence show that the root is 0.74 correct to 2 decimal places.

(a) $f(x) = \cos x - x$ so $f(0) = 1$ and $f\left(\frac{\pi}{3}\right) = 0.5 - 1.04... = -0.5...$

$f(0) > 0$ and $f\left(\frac{\pi}{3}\right) < 0$ therefore $\cos x - x = 0$ has a root between 0 and $\frac{\pi}{3}$

(b) Using $x_1 = 0.75$ and $x_{n+1} = \cos x_n$ gives

$$x_2 = \cos 0.75 = 0.73168...$$
$$x_3 = \cos 0.73168... = 0.74404...$$
$$x_4 = \cos 0.74404... = 0.73573...$$

Therefore $0.73573...$ is an approximate value of the root.

(c) $f(0.735) = \cos 0.735 - 0.735 = 0.0068... > 0$
$f(0.745) = \cos 0.745 - 0.745 = -0.0099... < 0$

Therefore the root, α, lies between 0.735 and 0.745, so $\alpha = 0.74$ correct to 2 decimal places.

Note that x is measured in *radians*, so the root is 0.74 rad.

 Exam tip

Iterations are easy to do on most scientific calculators: enter the value of x_1 and press EXE (or ENTER). Then enter the formula for $g(x_n)$ using ANS for each value of x. Then press EXE and continue to press EXE for each iteration.

Exercise 2.16

(a) Show that the equation $x^3 - 5x - 3 = 0$ has a root between -1 and 0.

(b) Use -0.5 as a first approximation for this root and the iteration given by $x_{n+1} = \dfrac{x_n^3 - 3}{5}$

Use six iterations to find a better approximation for the root, writing down 5 decimal places for each iteration.

(c) Show that your root is correct to 3 decimal places.

(d) The equation $x^3 - 5x - 3 = 0$ also has a root near $x = 2$
Explain why the iteration formula given will fail to find this root.

Section 2   Practice questions

1 A sequence is given by $u_1 = 8$ and $u_n = u_{n-1} - 2$
Show that the sequence is an arithmetic progression and write down the common difference.

2 The first three terms in a sequence are $\frac{a}{b}$, a and ab respectively, $b > 0$

(a) Show that the terms are in geometric progression.

(b) The first term is 2 and the product of the three terms is 216.
Find the values of a and b and the fifth term.

3 The nth term of a sequence, u_n, is given by
$$u_n = 2(3^n) - 4$$
Show that $u_{n+1} = 3u_n + 8$

4 The nth term of a sequence, u_n, is given by
$$u_n = \frac{2n^2 - n}{4n^2 + 1}$$
Show that the sequence converges and give the value to which it converges.

5 Determine whether the sequence whose nth term is $n \sin \frac{n\pi}{2}$ is alternating, periodic or oscillating.

6 Given $\displaystyle\sum_{r=2}^{r=n} u_r = \frac{n-1}{2n-1}$ find:

(a) u_n (b) $\displaystyle\sum_{r=n}^{r=2n} u_r$

(c) the sum to infinity of the series.

7 (a) Show that the terms of the series
$$\tfrac{3}{2} - \tfrac{1}{2} + \tfrac{1}{6} - \tfrac{1}{18} + \ldots$$
are in geometric progression.

(b) Find the sum of the first n terms of the series in (a).

(c) State with a reason whether the series is convergent.

8 (a) Express $\dfrac{1}{(r+1)(r+3)}$ in partial fractions.

(b) Hence find $\displaystyle\sum_{r=1}^{r=n} \frac{1}{(r+1)(r+3)}$

(c) Deduce the sum to infinity of the series
$$\frac{1}{2 \times 4} + \frac{1}{3 \times 5} + \frac{1}{4 \times 6} + \ldots$$

9 (a) Express $\dfrac{1}{(r-1)r(r+1)}$ in partial fractions.

(b) Hence find $\displaystyle\sum_{r=2}^{r=n} \frac{1}{(r-1)r(r+1)}$

10 The nth term of a sequence, u_n, is given by
$$u_n = n^2 - n$$
Prove by mathematical induction that the sum of the first n terms is given by $\frac{1}{3} n(n^2 - 1)$

11 $2000 is invested in an account that accrues interest at 5% per annum paid yearly. At the end of each year $500 is withdrawn from the account. Show that the amount A_n in the account after n years is given by
$$A_n = 2000(5 - 4(1.05)^n)$$

12 The rth term of a series, u_r, is given by
$$u_r = (2r - 1)(r + 2)$$
Find $\displaystyle\sum_{r=1}^{r=n} u_r$

13 (a) Use Maclaurin's theorem to find the first two terms in the expansion of
$$f(x) = e^x \sin x$$
as a series of ascending powers of x.

(b) Use your series to find an approximate value for $e^{\frac{\pi}{6}}$.

14 (a) Use Maclaurin's theorem to find the first four terms in the expansion of
$$\ln \sqrt{\frac{1-x}{1+x}}, \; -1 < x < 1$$
as a series of ascending powers of x, stating the values of x for which the expansion is valid.

(b) Use your series to find an approximate value for $\ln 3$.

15 (a) Find the first three terms of the expansion of $\cos x$ as ascending powers of $\left(x - \frac{\pi}{3}\right)$

(b) Hence find an approximate value of $\cos 61°$ given that $1° = 0.017\,\text{rad}$.

16 (a) Expand $\tan x$ as a series of ascending powers of $(x - a)$ as far as the term in $(x - a)^2$

(b) Use $a = \frac{\pi}{3}$ to find a quadratic function that gives an approximate value for $\tan x$ when x is close to $\frac{\pi}{3}$

17 (a) Prove that $^nC_r = {}^nC_{n-r}$

  (b) Find a relationship between n and r when
$$^nC_r = {}^{n+1}C_{r-1}$$

18 In the expansion of $\left(1 - \frac{1}{2}x\right)^9$ in ascending powers of x find:

  (a) the first four terms

  (b) the coefficient of x^7

  (c) the general term.

19 Find the terms up to and including x^2 in the expansion of
$$(1 - 2x)^4(1 + 4x)^6$$

20 Find the real part of $(1 - 2i)^5$

21 Find the term independent of x in the expansion of $\left(x^2 - \dfrac{2}{x}\right)^9$

22 Find the coefficient of x^4 in the expansion of
$$\frac{1}{\sqrt{2 - 3x}}$$

23 (a) Express $\dfrac{1}{(x + 1)(x + 2)(x^2 + 1)}$ in partial fractions.

  (b) Hence expand $\dfrac{1}{(x + 1)(x + 2)(x^2 + 1)}$ as a series of ascending powers of x up to and including the term in x^3, and give the range of values for which the expansion is valid.

24 Expand $(1 + x + 2x^2)^{-1}$ in ascending powers of x up to and including the term in x^3.

25 The diagram shows the graphs of the curves
$$y = e^{x-1} \text{ and } y = \frac{x + 2}{x}$$

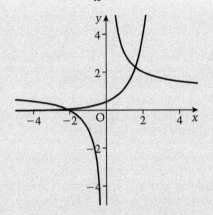

  (a) Verify that one solution of the equation
$$x e^{x-1} - x - 2 = 0$$
lies between $x = 1$ and $x = 2$

  (b) Use the interval bisection method twice to show that this root lies between $x = 1.75$ and $x = 2$

26 (a) Sketch the graphs of
$$y = x - 1 \text{ and } y = \ln(x + 2)$$
Use your sketch to show that the equation $1 - x + \ln(x + 2) = 0$ has only one positive root, α.

  (b) Use the intermediate value theorem to find two consecutive integers between which α lies.

  (c) Use linear interpolation twice to find an approximate value for α. Give your answer correct to 3 significant figures.

27 (a) Show that the equation $x^3 - 4x^2 + 5 = 0$ has a root between the turning points on the curve $y = x^3 - 4x^2 + 5$

  (b) Use the intermediate value theorem to find consecutive integers between which this root lies.

  (c) Use the Newton–Raphson method to find this root correct to 2 decimal places.

28 The diagram shows the curve $y = x^3 - 6x + 4$

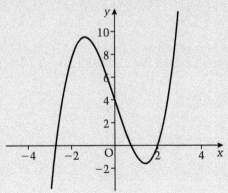

  (a) Confirm that the equation $x^3 - 6x + 4 = 0$ has one root equal to 2.

  (b) Using 1 as a first approximation to the other positive root, show that an iteration formula of the form
$$x_{n+1} = g(x_n)$$
converges to the value of this root and find it correct to 2 decimal places.

  (c) Taking the negative root as lying between -3 and -2, show that the same iteration formula fails to converge to this root.

  (d) Use another numerical method to find this root correct to 2 decimal places.

29 Use a numerical method to solve the equation
$$e^x = 3x$$
giving the roots correct to 3 decimal places.

3.1   The principles of counting

Counting

To answer any question starting 'How many … ?', we need an efficient method of counting.

When the entities to be counted can be placed in a one-to-one correspondence with the numerals, 1, 2, 3, … , counting them is easy. For example, to count the balls in a box, you can take them out one at a time counting 1, 2, 3, … as you go.

However, there are many situations where this is not possible. For example, how many different meals are possible when there is a choice of 3 main courses, 2 desserts and 2 drinks on a menu?

We can illustrate the different meals that can be chosen using a diagram:

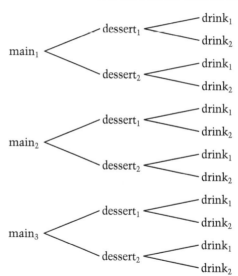

For each of the three ways of choosing a main course there are two ways of choosing a dessert.

Therefore there are 3×2 ways of choosing a main course and a dessert.

For each of these 3×2 ways there are two ways of choosing a drink.

Therefore there are $3 \times 2 \times 2$ different meals possible.

Now consider a multiple choice examination with 30 questions, each of which has a choice of four different possible answers. In how many different ways can this examination be answered?

Taking just the first two questions: for question 1 there are four different ways of choosing an answer and each of these four can be paired with one of the four different answers for question 2. This gives 4×4 different ways of answering the first two questions, i.e. 4^2 different ways.

Repeating this argument for all 30 questions gives 4^{30} different ways.

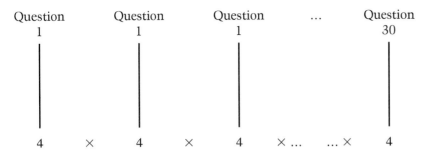

These two examples illustrate that

if there are *n* ways of doing one task, *m* ways of doing another, *l* ways
of doing yet another task ... and so on,
then the number of different ways of doing all the tasks is

$$n \times m \times l \times ...$$

This is known as the *fundamental counting principle*.

Example

Three ordinary six-sided dice, one red, one blue and one green, are rolled and a coin is tossed.
How many different outcomes are there?

There are six ways in which each dice can land and two ways in which the coin can land.

Therefore there are $6 \times 6 \times 6 \times 2 = 432$ different outcomes.

Example

A company selling software products uses a six-character code on each item.
The first character is one of the digits 1 to 9.
The next two characters are letters of the alphabet, not including vowels.
The next two characters are one of the digits 0 to 9.
The final character is a letter of the alphabet, not including the letter O.
How many different codes are possible when digits and letters can be repeated?

There are 9 choices for the first character, 21 choices for the next two characters, 10 choices for the fourth
and fifth characters and 25 choices for the last character.

Therefore there are $\quad 9 \times 21 \times 21 \times 10 \times 10 \times 25$
$$= 9\,922\,500 \text{ different codes.}$$

There are many other situations where we need an efficient method of
counting, and we look at some of them in the next few topics.

Exercise 3.1

1 There are three different colours of paper that can
be used to make a poster and there is a choice of
one of four different colours that can be used for
the print on the poster.

How many different colour combinations are
there?

2 The number plate on a car consists of three digits
followed by two letters of the alphabet, followed by
one digit. The first digit is 1 to 9, the next two digits
are 0 to 9, the two letters of the alphabet do not
include the letters I or O and the last digit is 1 or 0.

How many different number plates are possible
when digits and letters can be repeated?

3.2   Permutations

Learning outcomes

- To define a permutation and introduce the notation ^nP_r
- To find a variety of types of permutation

You need to know

- Factorial notation
- The fundamental principle of counting

Permutations

A *permutation* is an ordered arrangement of a number of objects.

For example, if four books, A, B, C and D, are placed on a shelf, one way of arranging them is A, B, C, D. Another is B, D, A, C.

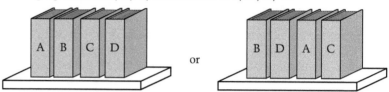

or

Each of these arrangements is called a permutation of the books and each arrangement is a different permutation.

The number of permutations is the number of different arrangements.

For the books, there are 4 different choices for the left-hand book. This leaves 3 different choices for the next book, so there are 4×3 different ways of selecting the first and second book. There are now only 2 ways of choosing the third book in the row, giving $4 \times 3 \times 2$ ways of arranging the first three books. There is only one book left, so the number of permutations of the four books is $4 \times 3 \times 2 \times 1 = 4!$

In general

the number of permutations of n different objects is $n!$

For example, the number of permutations of the 52 playing cards from an ordinary pack is $52!$

The examples of the books and cards are straightforward arrangements in a line. In the next example we look at the number of different arrangements of some of the n objects.

Example

How many different three-digit numbers can be made using the integers 2, 3, 4, 5, 6 if *each digit can only be used once*?

There are 5 ways of choosing the first integer, 4 ways of choosing the second integer and 3 ways of choosing the third integer.

Therefore there are $5 \times 4 \times 3 = 60$ different three-digit numbers that can be made.

The example is an illustration of a general case: the number of permutations of r objects from n different objects is
$n \times (n - 1) \times (n - 2) \times ... \times (n - r + 1)$

This can be written in factorial notation as $\dfrac{n!}{(n - r)!}$ and is denoted by ^nP_r.

i.e.   the number of permutations of r objects from n different objects is $^n\mathbf{P}_r = \dfrac{n!}{(n - r)!}$

In the next examples we look at arrangements that have conditions placed on them.

Example

How many different three-digit even numbers can be made using the integers 2, 3, 4, 5, 6 if *the digits can be repeated*?

The number has to be even so the last digit is restricted to 2, 4 or 6.

Starting from the right-hand end of the number, there are 3 different digits that can be used.

The next two digits can be any of the 5 given digits.

So there are $5 \times 5 \times 3 = 75$ different three-digit even numbers that can be made.

Circular arrangements

When n different objects are arranged in a circle, there is no first or last object.

For example, if five people, A, B, C, D and E, are to sit in chairs round a circular table, then there are 5 choices for chair 1, 4 choices for chair 2, and so on giving 5! different ways to be seated. This number includes the five arrangements shown in the diagrams:

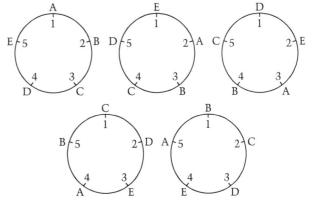

Now for any one of these arrangements, the people can be moved clockwise five times and each person will still have the same people on either side. Therefore the number of ways of seating the five people in numbered chairs is five times the number of ways of seating them round a circular table.

Therefore there are $\frac{5!}{5} = (5 - 1)!$ ways of arranging five different objects in a circle.

In general

there are $(n - 1)!$ ways of arranging n different objects in a circle and $\frac{n!}{r(n - r)!}$ ways of arranging r objects from n different objects in a circle.

Now consider the number of arrangements of five different beads on a circular ring.

The $\frac{5!}{5}$ different arrangements include these two:

A ring can be turned over, so these two arrangements are the same.

Therefore the number of different arrangements in a ring is half the number of different arrangements in a circle. So there are $\frac{5!}{2 \times 5} = 12$ ways of arranging the five beads on a ring.

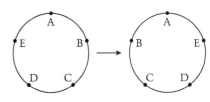

In general

when n different objects are arranged in a ring that can be turned over there are $\frac{n!}{2n} = \frac{1}{2}(n - 1)!$ different ways of doing this.

Exercise 3.2a

1 In how many different ways can the letters in the word PAGES be arranged?

2 How many three-digit numbers can be made from the digits 3, 5, 6 and 7 if

(a) the number is odd and each digit can be used once

(b) the number is even and each digit can be used more than once?

3 In how many different ways can three beads from five different beads be threaded on a ring?

Permutations when not all the objects are different

Consider the number of different ways of arranging the letters in the word LOOK.

There are two letters O in this word. If we label them as O_1 and O_2 then the number of different arrangements of the letters L O_1 O_2 K is 4!

But this number includes the two arrangements

L O_1 O_2 K   and   L O_2 O_1 K

so the arrangement L O O K appears twice in the 4! number.

This means that the number of arrangements of the letters L O O K is
$$\frac{4!}{2!} = 12$$

Applying the same argument to the letters in the word CURRICULUM, we have two Cs, two Rs and three Us, so the number of arrangements of the letters C_1 U_1 R_1 R_2 I C_2 U_2 L U_3 M is 10!

But 10! includes the 2! ways of arranging the two Cs, the 2! ways of arranging the two Rs and the 3! ways of arranging the three Us.

Therefore the number of arrangements of the letters in CURRICULUM is $\frac{10!}{2!2!3!} = 151\,200$

In general

> **the number of permutations of n objects when p are the same and q are the same is $\dfrac{n!}{p!q!}$**

Permutations when some objects have to be kept together or kept apart

To find the number of permutations of the letters in the word THREE when the two Es are kept together, we can consider the two Es as *one object*, i.e. find the number of permutations of the *four* objects T H R (EE), which is 4!

This means that the number of permutations where the two Es are apart is the total number of permutations of the letters minus the number where the Es are together,

i.e. $\dfrac{5!}{2!} - 4! = 36$

Independent permutations

Two tasks are **independent** when the execution of one task has no effect on the execution of the other task.

For example, the number of different number plates with any two letters followed by any four digits is the number of permutations of two letters of the alphabet, $^{26}P_2$, and the number of permutations of four digits, $^{10}P_4$. These two permutations have no effect on each other, so the permutations are independent. Using the fundamental principle of counting, the number of different number plates is $^{26}P_2 \times {}^{10}P_4$

> **Therefore when two tasks are independent, the number of ways of doing both is the product of the number of ways of doing each task.**

Mutually exclusive permutations

Two tasks are ***mutually exclusive*** when they cannot both be executed.

For example, it is impossible to make a two-digit number and a three-digit number – a number either has two digits or it has three digits, not both.

Using the digits 1, 2, 3, 4, 5 without repeating a digit,

the number of permutations giving a two-digit number is $\frac{5!}{3!} = 20$

the number of ways of making a three-digit number is $\frac{5!}{2!} = 60$

and these two permutations cover all the different two-digit numbers and three-digit numbers so there are $20 + 60 = 80$ ways of making a two-digit number or a three-digit number from 1, 2, 3, 4, 5 without repeating a digit.

> **When two tasks are mutually exclusive, the number of ways
> of doing either one task or the other is the sum of the
> number of ways of doing each task.**

Example

A number plate with five characters on it consists of at least three letters together followed by at least one digit. The letters are chosen without repetition from the letters A, B, C, D, E, F and the digits are selected without repetition from the digits 1 to 9 inclusive.
The letters must include the letter A. Find the number of different number plates possible.

There must be either 3 letters and 2 digits or 4 letters and 1 digit:

- 3 letters and 2 digits
  There are 2 letters available from the 5 remaining letters (B, C, D, E, F) and 2 digits from the 9 digits. There are 5×4 ways of arranging the two letters and for each of these there are 3 positions that A can be in. So there are $3 \times 5 \times 4$ permutations of the letters. There are 9×8 permutations of the digits. The permutations of letters and digits are independent, so the number of permutations of 3 letters and 2 digits is $3 \times 5 \times 4 \times 9 \times 8 = 4320$

- 4 letters and 1 digit
  The number of permutations of letters including A (using similar reasoning to the first case) is $5 \times 4 \times 3 \times 4$ and the number of permutations of one digit is 9. Therefore the number of permutations of 4 letters and 1 digit is $5 \times 4 \times 3 \times 4 \times 9 = 2160$

The two cases considered are mutually exclusive, so the number of different number plates is
$4320 + 2160 = 6480$

Exercise 3.2b

1 Find the number of arrangements of the letters in the word PROBABILITY in which
   (a) the Bs are together
   (b) the Is are apart
   (c) the Bs are together and the Is are apart.

2 Three multiple choice questions each have one correct answer and three incorrect answers.
   In how many ways can these questions be answered so that there is at least one correct answer?

3 A code is made from three digits selected from the digits 1, 2, 3, 4, 5, 6.
   In how many of these codes are the digits in ascending order of size?

Combinations

Learning outcomes

- To define a combination
- To find a variety of combinations
- To distinguish between a permutation and a combination

You need to know

- How to find a permutation
- What independent permutations are
- What mutually exclusive permutations are
- The meaning of ^nC_r

Combinations

We have seen that the number of different arrangements of 4 books on a shelf is 4!, but there is only one set, or combination, of books.

A **combination** is a group of objects when the order of the objects in the group does not matter.

Suppose we want to find how many groups of 5 books can be selected from 8 different books.

There are $^8P_5$ different arrangements of 5 books selected from the 8 books, but this number of arrangements includes the 5! arrangements of the 5 books selected among themselves.

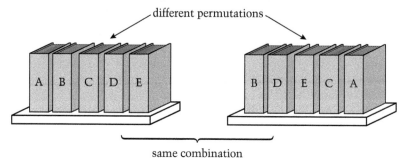

different permutations

A | B | C | D | E   B | D | E | C | A

same combination

Therefore the number of different combinations of 5 books selected from 8 different books is

$$\frac{^8P_5}{5!} = \frac{8!}{5!(8-5)!}$$

Now $\dfrac{8!}{5!(8-5)!} = {}^8C_5$ so we can denote the number of combinations of objects chosen from 8 different objects by $^8C_5$.

The same argument applies to the general case:

the number of different combinations of r objects selected from n different objects

$$\textbf{is given by } \quad {}^nC_r = \frac{n!}{r!(n-r)!}$$

For example, the number of different ways of selecting 8 people from a group of 10 people is

$$^{10}C_8 = \frac{10!}{8!(10-8)!} = 45$$

Example

In how many ways can a set of 8 students be divided into two equal groups?

There must be 4 students in each group.

The number of ways of selecting 4 students from 8 is $^8C_4$ and this leaves the remaining students as the other group.

Labelling the students A, B, C, D, E, F, G, H, one selection is the group (A, B, C, D).

This gives (A, B, C, D) and (E, F, G, H) as the two groups. But (E, F, G, H) is one of the selections included in the $^8C_4$ selections and this gives (E, F, G, H) and (A, B, C, D) as the two groups. So $^8C_4$ gives twice the number of divisions into two equal groups.

Therefore the number of ways the students can be divided into two equal groups is

$$\frac{^8C_4}{2} = \frac{8!}{2 \times 4! \times 4!} = 35$$

Exercise 3.3a

1 A box holds a large number of red, blue, yellow, green, black and brown balls. How many selections of four balls can be made if

   (a) they are all different colours

   (b) two balls only are the same colour?

2 In how many ways can 10 children be divided into two groups of 6 children and 4 children?

Distinguishing between permutations and combinations

Problems do not usually include the words permutation or combination. You need to read the problem carefully and use the context to decide whether or not the order of any selection matters.

For example, the number of different hands of cards that can be dealt from an ordinary pack of 52 playing cards is a number of combinations as a hand of cards is a group and the order does not matter. The number of different numbers than can be made from 4 of 5 different digits is a permutation as the order of digits in a number does matter.

Example

How many different ways can 4 students be selected from 10 students if either Martha or Sergio but not both must be selected?

Either   Martha is selected and Sergio is not, which leaves 3 students to be selected from the remaining 8 students, giving $^8C_3$ different groups,

or   Sergio is selected and Martha is not, which again leaves 3 students to be selected from the remaining 8 students, giving $^8C_3$ different groups.

These are mutually exclusive combinations so the number of different groups is $2\,^8C_3 = 112$

Exercise 3.3b

1 Find the number of ways in which 10 girls can be placed in a line so that Alice, Grace and Maria are separated.

2 The diagram shows a grid of 8 vertical lines and 7 horizontal lines.

   Starting at A, and either moving left or up at each intersection, how many routes are there to get to B?

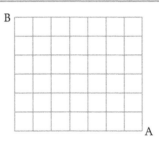

Sample spaces

When we perform a task, one of the items that results from the task is called an **outcome**.

For example, if the task is choosing two letters from the letters A, B, C, D, one possible outcome is AB.

All the possible outcomes of a task is called a **sample space**.

Tables

When a task involves just two items, drawing up a **two-way table** is a good method for ensuring that all the possible outcomes are listed.

For example, this table lists all the possible outcomes when two letters are chosen from A, B, C and D.

	A	B	C	D
A	AA	AB	AC	AD
B	BA	BB	BC	BD
C	CA	CB	CC	CD
D	DA	DB	DC	DD

The possible choices of letter are listed along the top and down the left-hand side of the table. Then the table can be filled in with the outcomes.

This table gives all the outcomes when a coin is tossed and a six-sided dice is rolled. The table shows, for example, that there are two outcomes resulting in a head and a score greater than 4.

	1	2	3	4	5	6
H	H1	H2	H3	H4	H5	H6
T	T1	T2	T3	T4	T5	T6

Tables are also useful when a sample space contains only a few different outcomes but each outcome occurs several times.

For example, this table shows the outcomes and the number of times each outcome occurred in a drug trial for a new treatment for migraine.

Gender	Big improvement		Mild improvement		No improvement	
	No side-effects	Some side-effects	No side-effects	Some side-effects	No side-effects	Some side-effects
Male	25	3	65	2	38	1
Female	21	5	97	1	14	0

From this we can read, for example, that the number of outcomes giving a big improvement is 54 in total.

Tree diagrams

Tables are not suitable when there are several different outcomes for a task involving more than two items, for example when three different letters, chosen from the letters A, B, C, D, are arranged in a line. We know that there are $^4P_3 = 24$ different outcomes, but not what these outcomes are. We can find these outcomes by drawing a *tree diagram*:

Start by drawing 4 branches to show the 4 different choices for the first letter, writing the letter on each branch	At the end of each branch, repeat for the 3 different choices of the second letter	Repeat for the 2 different choices for the third letter	At the end of the branch read along the path to list the outcome

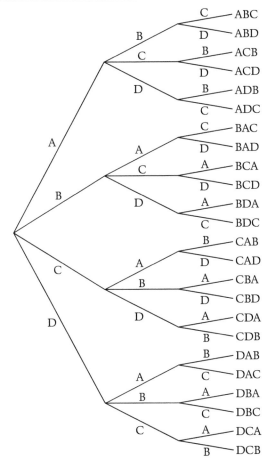

The list at the right-hand end of the diagram is the sample space for this task.

Venn diagrams

When a task involves overlapping outcomes, we can sometimes use a *Venn diagram* to illustrate and find the numbers of the different outcomes.

Consider, for example, this information for 50 students from a college entered for CAPE examinations:

- 10 students entered for physics, P
- 12 students entered for chemistry, C
- 5 students entered for both physics and chemistry.

This does not give the numbers of students who entered for physics but not for chemistry, or vice-versa. It does tell us how many entered for physics and chemistry, so there is an overlap between the numbers entered for physics and the number entered for both subjects.

We can represent this on a Venn diagram using overlapping circles to represent the numbers entered both for P and for C. In the overlap region, we enter 5, which leaves 5 in the non-overlapping part of the circle P and 7 in the non-overlapping part of the circle C.

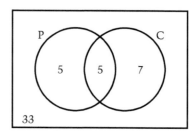

Therefore 17 students were entered for physics or chemistry or both, leaving 33 who were entered for neither subject. This number goes in the box outside the circles.

Example

Of the 100 students in a school entered for CSEC examinations:

45 were entered for mathematics

25 were entered for geography

18 were entered for mathematics and geography

36 were entered for none of these subjects.

30 were entered for economics

15 were entered for mathematics and economics

15 were entered for economics and geography

(a) Draw a Venn diagram to show this information.

(b) Find the number of students who
(i) entered for all of these subjects   (ii) entered for mathematics but not economics nor geography.

This needs a Venn diagram with three overlapping circles. We will use M for mathematics, E for economics and G for geography to label the circles containing the numbers for each subject. We need regions where just M and E overlap, just M and G overlap, just E and G overlap and where all three overlap.

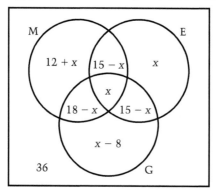

(a) We do not know how many students entered for all three subjects so we put x in the region where all three circles overlap.

Looking at the numbers for mathematics and geography, we know that 18 students entered both. The region where M and G overlap already contains x, so that leaves $18 - x$ for the region where just M and G overlap.

Now looking at the numbers for mathematics and economics, 15 enter for both. Taking out the number who enter all three subjects leaves $15 - x$ for the region where just M and E overlap.

We know that 45 is the total number in M, so that leaves $(45 - (33 - x)) = 12 + x$ in the region where M does not overlap either of the other circles.

The numbers in the remaining regions can be filled in using similar reasoning,

(b) (i) The sum of the numbers in all the regions is $88 + x$
There are 100 students in total, therefore $88 + x = 100 \Rightarrow x = 12$
Therefore 12 students were entered for all three of these subjects.

(ii) Reading from the diagram, 24 students entered for mathematics but not economics nor geography.

Exercise 3.4

1 An ordinary six-sided red dice and an ordinary six-sided blue dice are both rolled.
   (a) How many different outcomes are there?
   (b) Draw a table showing the sample space.

2 The table shows the outcomes of an investigation into the age of cars owned.

Age of car	Over 10 years				5–10 years			
Age of owner	18–25	26–40	41–60	over 60	18–25	26–40	41–60	over 60
	40	35	15	30	55	60	32	65

Age of car	Under 5 years			
Age of owner	18–25	26–40	41–60	over 60
	5	8	56	20

   (a) How many people owned a car over 10 years old?
   (b) How many people up to the age of 60 own a car that is 10 years old or less?

3 Four different coins are tossed at the same time.
   (a) Draw a tree diagram to show all the outcomes.
   (b) How many outcomes result in at least two heads?

4 Of the 100 customers at a market stall selling vegetables:
   37 bought sweet potatoes
   28 bought tomatoes
   56 bought carrots
   15 bought sweet potatoes and tomatoes
   12 bought tomatoes and carrots
   16 bought sweet potatoes and carrots
   12 did not buy sweet potatoes, tomatoes or carrots.

   (a) Draw a Venn diagram to show this information.

   (b) (i) How many people bought sweet potatoes, tomatoes and carrots?
   (ii) How many people bought just carrots?

Learning outcomes

- To introduce the terminology used in probability
- To define and use basic probability

You need to know

- How to find permutations and combinations
- How to read tables and Venn diagrams
- What outcomes and sample spaces mean

Terminology

Probability gives a measure for how likely it is that an event will happen, i.e. probability gives a measure of predictability.

Up to now we have talked about tasks, but in the context of probability we call tasks *experiments*. For example, choosing three letters from A, B, C and D is called an experiment.

An *event* is an outcome or a group of outcomes from an experiment. For example, the outcome ABC is an event when choosing three letters from A, B, C and D. An event can also be all the outcomes containing the letter A.

When one letter is selected from A, B, C and D, then the selection is *random* when the selection of any one letter is as likely as the selection of any other letter. In this case, we can say that each outcome is *equally likely*.

When coins or dice are involved in experiments, they are described as *fair* or *unbiased* if the coins are equally likely to land heads up or tails up and if the dice are equally likely to show any one of the possible scores.

Definition of probability

When all the outcomes of an experiment are known the probability that an event A is likely to happen is denoted by $P(A)$ and is given by

$$P(A) = \frac{\text{the number of equally likely outcomes giving } A}{\text{the total number of equally likely outcomes}}$$

Depending on the nature of the event A, the numerator of this fraction can be any number from zero (no outcomes giving A) to the number in the denominator (all equally likely outcomes giving A).

$$\text{Therefore} \quad 0 \leqslant P(A) \leqslant 1$$

When $P(A) = 0$, the event A is impossible and when $P(A) = 1$, the event is certain to happen.

Probabilities are given as fractions or decimals or percentages.

For example, when an ordinary fair dice is rolled, each of the 6 scores is equally likely.

To find the probability that the score will be greater than 4, we know that the number of equally likely outcomes is 6 and the event 'a score of 5 or 6' is 2 of the equally likely outcomes.

Therefore $P(5 \text{ or } 6) = \frac{2}{6} = \frac{1}{3}$

Also $P(\text{score is } 7) = 0$ and $P(\text{score is 6 or less}) = 1$

Example

The table gives a breakdown of car theft in an island for the year 2012.

Cost of a replacement	Age of stolen car in years		
	Less than 1	1−3	Older than 3
Less than $10 000	22	30	60
$10 000−$30 000	78	56	84
More than $30 000	14	25	8

Find the probability that a randomly selected theft was

(a) of a car up to 3 years old

(b) of a car older than one year and costing $10 000 or more to replace?

(a) There are 377 thefts listed in the table and any one of these is equally likely to be selected.
Thefts of cars up to 3 years old are listed in the first two columns: there 225 of these.

$$\therefore \quad P(\text{theft was a car up to 3 years old}) = \frac{225}{377} = 0.597 \ (3 \text{ s.f.})$$

(b) Let B be the event 'thefts of cars older than one year and costing $10 000 or more to replace'.
These are listed in the lower two right-hand columns and rows of the table: there are 173.

$$\therefore \quad P(B) = \frac{173}{377} = 0.459 \ (3 \text{ s.f.})$$

Example

Two cards are drawn at random from a pack of 20 cards containing 5 red cards, 5 blue cards, 5 yellow cards and 5 green cards. Find the probability that both cards are red.

The number of combinations of two red cards is $^5C_2$ and the number of combinations of any two cards is $^{20}C_2$.

Therefore the probability that two red cards are drawn is

$$\frac{^5C_2}{^{20}C_2} = \frac{5 \times 4}{20 \times 19} = 0.0526 \ (3 \text{ s.f.})$$

Exercise 3.5

1   Two digits are selected at random from the digits 1, 2, 3, 4, 5, 6, 7 to make a two-digit number. What is the probability that this number

   (a) is even

   (b) contains two odd digits?

2   Three different letters, chosen at random from the letters A, B, C, D, are arranged in a line. Using the tree diagram in Topic 3.4 or otherwise, find the probability that the letters A and B are next to each other.

3   A box contains 200 different patterned tiles of mixed colours on a white background.

   In the pattern, 65 tiles include red, 39 tiles include blue, 53 tiles include yellow, 20 tiles include red and blue, 18 tiles include red and yellow, 25 tiles include blue and yellow and 10 tiles include all three colours.

   (a) Draw a Venn diagram to show this information.

   (b) One tile is selected at random. Find the probability that the pattern on it contains
      (i) only red
      (ii) red and blue but not yellow.

Learning outcomes

- Finding the probability that an event does not happen

You need to know

- Simple set notation
- What a sample space is
- How to use basic probability
- The meaning of mutually exclusive

The probability that an event does not happen

If A is an event, then the event 'not A' is denoted by A'.

When an ordinary six-sided dice is rolled, the sample space is the set $S = \{1, 2, 3, 4, 5, 6\}$.

S contains 6 elements so $n(S) = 6$

This sample space contains every possible outcome so it is **exhaustive**.

If the dice is unbiased, the outcomes are all equally likely, so if A is the event of scoring 1 or 2,

then the number of ways in which A can occur is 2, so $P(A) = \frac{2}{6}$

Now the number of ways in which A cannot occur is $6 - 2$,

Therefore $\quad P(A') = \dfrac{6 - 2}{6} = \dfrac{6}{6} - \dfrac{2}{6} = 1 - P(A)$

In general if the number of equally likely ways an event A can happen is x and the sample space is S, then the number of ways in which A cannot happen is $n(S) - x$,

$$\therefore \quad P(A') = \frac{n(S) - x}{n(S)} = \frac{n(S)}{n(S)} - \frac{x}{n(S)} = 1 - P(A)$$

$$\textbf{i.e.} \quad \textbf{P}(A') = \textbf{1} - \textbf{P}(A)$$

For example, the probability that an unbiased dice shows 6 when rolled is $\frac{1}{6}$, therefore the probability that the dice does not show 6 is $1 - \frac{1}{6} = \frac{5}{6}$

and if the probability that it will rain tomorrow is 67%, then the probability that it will not rain tomorrow is $100\% - 67\% = 33\%$

In simple cases, $P(A')$ can be found directly, for example when one letter is chosen at random from the letters A, B, C, D, the probability that it is not the letter D is $\frac{3}{4}$

In other cases it may be easier to find $P(A)$ first.

Example

A two-digit number, greater than zero, is made by choosing two integers at random from the digits 0 to 9 inclusive. A digit can be chosen more than once. What is the probability that the number is not a multiple of 5?

It is easier to find how many numbers are multiples of 5 than how many are not.

A number is divisible by 5 if it ends in 0 or 5.

The number of permutations of two digits ending in 0 or 5 is 9×2

The number of permutations of two digits is 9×10

A number cannot start with 0 but it can end with 0

So if A is 'the integer is a multiple of 5'

$$P(A) = \frac{9 \times 2}{9 \times 10} = \frac{1}{5}, \quad \therefore \quad P(A') = 1 - P(A) = 1 - \frac{1}{5} = \frac{4}{5}$$

Example

This Venn diagram (from Topic 3.4) shows, out of the 100 students who were entered for CSEC examinations, the numbers who were entered for mathematics, economics and geography.

One of these 100 students is chosen at random. What is the probability that they were entered for at least one of the subjects mathematics, economics or geography?

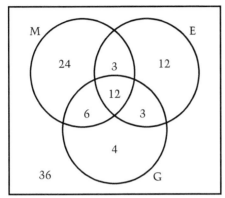

It is easier to find the probability that a student was not entered for at least one of the subjects. The Venn diagram shows that this number is 36.

Taking A as the event 'a student is entered for at least one of the subjects mathematics, economics or geography', A' is the event 'a student is not entered for any of the subjects mathematics, economics or geography'.

Then $P(A') = \dfrac{36}{100} = 0.36$

Using $P(A') = 1 - P(A)$ gives $0.36 = 1 - P(A)$

$\therefore \qquad P(A) = 0.64$

Exercise 3.6

1 Two pens are chosen at random from a box containing 6 red, 4 blue and 8 black pens.
  What is the probability that at least one pen is blue?

2 This table from Topic 3.4 shows the outcomes for 272 people taking part in a drug trial for a new treatment for migraine.

Gender	Big improvement		Mild improvement		No improvement	
	No side-effects	Some side-effects	No side-effects	Some side-effects	No side-effects	Some side-effects
Male	25	3	65	2	38	1
Female	21	5	97	1	14	0

Find the probability that one person, chosen at random, had no side-effects.

Learning outcomes

- To find the probability that events A and B will both occur
- To find the probability that either event A will occur or event B will occur

You need to know

- How to use set notation
- The meaning of union and intersection of sets
- Basic probability
- About independent permutations and combinations

Mutually exclusive events

Two events are ***mutually exclusive*** when they cannot both occur simultaneously.

For example, choosing an even number and choosing an odd number are mutually exclusive because a number cannot be both even and odd.

However, choosing a number that is a multiple of 3 and choosing any even number are not mutually exclusive because 6, for example, is both.

If A and B are mutually exclusive events, the set of ways in which A can occur and the set of ways in which B can occur will have no members in common. They can be represented in a Venn diagram as two circles that do not overlap. The set S is the sample space.

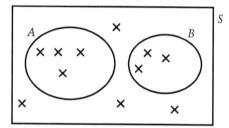

If there are n equally likely outcomes in the sample space of which A can occur in p ways and B can occur in q ways,

then the probability of A or B occurring is $\dfrac{p + q}{n} = \dfrac{p}{n} + \dfrac{q}{n} = P(A) + P(B)$

The probability that A or B will occur is denoted by $P(A \cup B)$.

Therefore when A and B are mutually exclusive events
$$P(A \cup B) = P(A) + P(B)$$

Example

The probability that a girl walks to school is $\frac{1}{4}$ and the probability that she takes a bus to school is $\frac{3}{5}$. What is the probability that she goes to school by another method?

P(she goes to school by another method) = P(she does not walk nor take a bus)
                                   = 1 − P(she does walk or take a bus)

Walking and taking a bus are mutually exclusive, therefore

P(she does walk or take a bus) $= \frac{1}{4} + \frac{3}{5} = \frac{17}{20}$

∴   P(she goes to school by another method) $= 1 - \frac{17}{20} = \frac{3}{20}$

Independent events

Two events are ***independent*** when one event, whether or not it occurs, has no effect on whether or not the other event occurs.

For example, rolling an ordinary fair six-sided dice and tossing a fair coin are independent experiments.

The number of ways the dice can land and the coin can land is 6×2

The number of ways the dice can show five uppermost and the coin can show a head is 1×1

Therefore $P(5 \text{ and } H) = \frac{1 \times 1}{6 \times 2} = \frac{1}{6} \times \frac{1}{2} = P(5) \times P(H)$

If there are n outcomes in an experiment in which an event A can occur in p ways,

then $P(A) = \frac{p}{n}$

If there are m outcomes in an independent experiment in which event B

can occur in q ways, then $P(B) = \frac{q}{m}$

As the experiments are independent, A and B are independent.

There are $n \times m$ outcomes for the first *and* second experiment in which A *and* B can occur in p and q ways,

so $P(A \text{ and } B) = \frac{p \times q}{n \times m} = \frac{p}{n} \times \frac{q}{m} = P(A) \times P(B)$

$P(A \text{ and } B)$ is denoted by $P(A \cap B)$.

Therefore when A and B are independent events
$$P(A \cap B) = P(A) \times P(B)$$

Example

Three ordinary six-sided dice are rolled together. Two of the dice are fair but the third dice is biased so that it is twice as likely to show 6 as any other number. Find the probability that all three dice will show 6.

The equally likely outcomes from the biased dice are 1, 2, 3, 4, 5, 6, 6. Therefore the probability that this dice shows 6 is $\frac{2}{7}$. The probability that a fair dice shows 6 is $\frac{1}{6}$

There are three dice, so we will call the fair dice a and b and the biased dice c and events of showing a 6 as, 6_a, 6_b and 6_c.

The way in which one of the dice lands has no effect on the way the others land, so the events are independent.

$\therefore \quad P(6_a \cap 6_b \cap 6_c) = P(6_a) \times P(6_b) \times P(6_c)$
$$= \frac{1}{6} \times \frac{1}{6} \times \frac{2}{7} = \frac{2}{252}$$

Probability that *A* or *B* occurs when *A* and *B* are not mutually exclusive

Suppose that one number is chosen at random from the integers 1 to 12 inclusive and we want to find the probability that the number is even or a multiple of 3.

If A is the event 'choosing an even number' and B is the event 'choosing a multiple of 3' then A and B are not mutually exclusive because 6 and 12 are both even and multiples of 3.

We can illustrate the sample space in a Venn diagram.

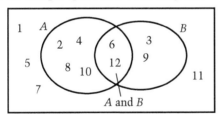

This diagram shows that the number of elements in either A or B, i.e. in $A \cup B$, is not (number of elements in A) + (number of elements in B) because this includes the elements 6, 12, i.e. in $A \cap B$, twice.

The number of ways in which A can occur is 6 so $P(A) = \dfrac{6}{12}$

The number of ways in which B can occur is 4 so $P(B) = \dfrac{4}{12}$

The number of ways in which A and B can occur is 2, so $P(A \cap B) = \dfrac{2}{12}$

The number of ways in which A or B can occur is $8 = 6 + 4 - 2$,

so $P(A \cup B) = \dfrac{6 + 4 - 2}{12} = \dfrac{6}{12} + \dfrac{4}{12} - \dfrac{2}{12} = P(A) + P(B) - P(A \cap B)$

For any two events A and B that are not mutually exclusive:

(the number of ways in which A or B can occur)

= (number of ways in which A can occur) + (number of ways in which B can occur) − (number of ways in which A and B can occur).

This is because (number of ways in which A can occur) + (number of ways in which B can occur) includes (number of ways in which A and B can occur) twice.

Therefore when A and B are not mutually exclusive events
$$P(A \cup B) = P(A) + P(B) - P(A \cap B)$$

Notice that the set of elements in the part of B that excludes $\{A \cap B\}$ is described by the set $\{A' \cap B\}$ and similarly for the set of elements in the part of A that excludes $\{A \cap B\}$

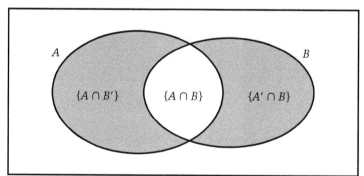

Example

This table from Topic 3.4 shows the outcomes for 272 people taking part in a drug trial for a new treatment for migraine.

Gender	Big improvement		Mild improvement		No improvement	
	No side-effects	Some side-effects	No side-effects	Some side-effects	No side-effects	Some side-effects
Male	25	3	65	2	38	1
Female	21	5	97	1	14	0

Find the probability that one person, chosen at random, is male or has side-effects.

Being male and having side-effects are not mutually exclusive. Therefore

P(male *or* side-effects) = P(male) + P(side-effects) − P(male *and* side-effects)

The number of {males} is 134, the number of {side-effects} is 12,

the number of {male *and* side-effects} is 6.

Therefore P(male *or* no side-effects) $= \dfrac{134}{272} + \dfrac{12}{272} - \dfrac{6}{272}$

$$= \dfrac{140}{272} = 0.515 \ (3 \text{ s.f.})$$

Example

Two fair normal six-sided dice are rolled together. Find the probability of rolling at least one 6 or at least one 5.

At least one 6 and at least one 5 are not mutually exclusive, therefore

P(at least one 6 *or* at least one 5)

$\qquad$ = P(at least one 6) + P(at least one 5) − P(at least one 6 *and* at least one 5)

$\qquad$ P(at least one 6) $= 1 - \text{P(no 6)} = 1 - \dfrac{5 \times 5}{36} = \dfrac{11}{36}$

$\qquad$ P(at least one 5) $= 1 - \text{P(no 5)} = 1 - \dfrac{5 \times 5}{36} = \dfrac{11}{36}$

$\qquad$ P(at least one 6 *and* at least one 5) $= \dfrac{2 \times 2}{36} = \dfrac{4}{36}$

$\therefore \quad$ P(at least one 6 *or* at least one 5) $= \dfrac{22 - 4}{36} = \dfrac{1}{2}$

Exercise 3.7

1 Two boys, A and B play a game that involves rolling an ordinary six-sided dice.
The first person to roll a six wins. A goes first. Find the probability that B wins on his first turn.
(Hint: For B to win on his first turn, A must lose on his first turn.)

2 A and B are independent events. $P(A) = \frac{2}{5}$, $P(B) = \frac{1}{4}$ and $P(A' \cap B) = \frac{1}{10}$

Find **(a)** $P(A \cap B)$ $\qquad\qquad$ **(b)** $P(A \cup B)$

3 A tennis player A has a probability of $\frac{3}{5}$ of winning a set against player B. The first player to win 2 sets out of 3 wins the match.
Find the probability that when A plays B, A wins the match.

4 A and B are two events such that $P(A) = 0.3$, $P(B) = 0.5$ and $P(A \cup B) = 0.6$

Explain why A and B are neither mutually exclusive nor independent.

Events that are not independent

If a card is removed at random from a pack of 4 cards numbered 1, 2, 2, 3 and not put back, and then a second card is chosen, the options for the number on the second card depend on which number was removed first.

Therefore these two events are not independent as the ways in which the second event can occur have been reduced by one and depend on which number was removed first.

The number of equally likely outcomes is therefore $4 \times 3 = 12$

We can show the probabilities for each card and the different outcomes on a tree diagram:

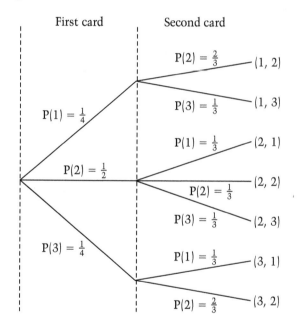

Notice that this tree diagram shows the seven different outcomes, which are mutually exclusive, but it does *not* show all the equally likely outcomes (there are 12).

For example, there are two ways in which the first card is 1 and the second card is 2, i.e. $(1, 2_1)$ and $(1, 2_2)$

so   P(1st card is 1 *and* 2nd card is 2) $= \frac{2}{12} = \frac{1}{6}$

Using the tree diagram,

     P(1st card is 1) $\times$ P(2nd card is 2) $= \frac{1}{4} \times \frac{2}{3} = \frac{1}{6}$

Similarly,

     P(1st card is 2 *and* 2nd card is 1) $= \frac{1}{6} =$ P(1st card is 2) $\times$ P(2nd card is 1)

So, to find the probability of an outcome shown on a tree diagram, we multiply the probabilities on the path leading to that outcome.

There are two outcomes where the two cards are 1 and 2 in any order, i.e. (1, 2) and (2, 1).

These are mutually exclusive, so to find the probability that one card is 1 and the other is 2, we add the probabilities of each.

Therefore P(the cards removed are numbered 1 and 2) $= \frac{1}{6} + \frac{1}{6} = \frac{1}{3}$

To find the probability of two or more outcomes shown on a tree diagram, we add the probabilities of each outcome.

Tree diagrams can be extended to cover more than two events.

This diagram shows the probabilities when three coins are tossed, two of which are fair and one is biased so that a head is twice as likely as a tail.

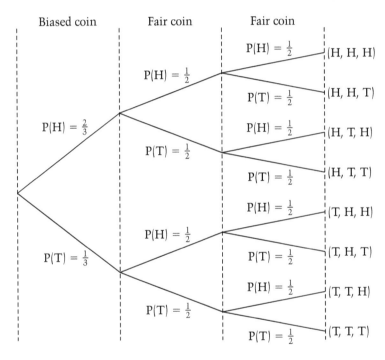

To find the probability of any one of the outcomes, we multiply the probabilities along the path giving that outcome.

For example, to find the probability that the biased coin shows a head, and the other two coins show tails, we follow the path leading to (H, T, T) giving P(H, T, T) $= \frac{2}{3} \times \frac{1}{2} \times \frac{1}{2} = \frac{1}{6}$

To find the probability of more than one outcome we add the probabilities of each.

For example, to find the probability that the three coins land showing two heads and a tail, we add the probabilities of the events
(H, H, T), (H, T, H) and (T, H, H) giving $\frac{1}{6} + \frac{1}{6} + \frac{1}{12} = \frac{5}{12}$

There are problems involving two events when we are interested in only some of the possible outcomes. In cases like these we can draw a simplified tree diagram.

Example

Three ordinary six-sided dice are rolled together. Two of the dice are fair and one is biased so that a six is three times as likely as any other score.

Find the probability that exactly two sixes are rolled.

We are interested in whether sixes are rolled or not, so we need only show probabilities and outcomes for sixes or not sixes.

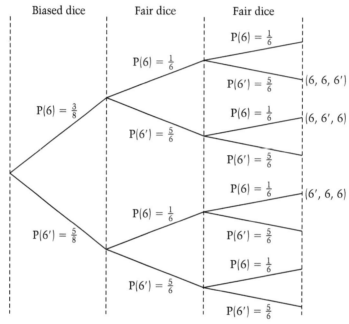

Therefore P(two 6s) $= \left(\frac{3}{8} \times \frac{1}{6} \times \frac{5}{6}\right) + \left(\frac{3}{8} \times \frac{5}{6} \times \frac{1}{6}\right) + \left(\frac{5}{8} \times \frac{1}{6} \times \frac{1}{6}\right) = \frac{35}{288} = 0.122$

Exercise 3.8a

1 A pack of ten cards are numbered 1 to 10. One card is removed at random and not replaced. A second card is then removed at random. Find the probability that the sum of the numbers on the two cards is 3.

2 A bag contains 3 red pens and 2 blue pens. One pen is removed at random and not replaced, then a second pen is removed. Find the probability that one red and one blue pen are removed.

Conditional probability

We refer to the card situation outlined at the start of this topic, namely, a card is removed at random from a pack of 4 cards numbered 1, 2, 2, 3 and not put back, and then a second card is removed at random. Suppose we want to find the probability that the second card removed is number 2, given that the first card is 1.

This is an example of *conditional probability* and we write it as
P(2nd card is 2 | 1st card is 1)

In general $P(A|B)$ means the probability that A occurs given that B has already occurred.

If the first card is numbered 1, there are 2 out of 3 ways of removing a card numbered 2, so

$$P(\text{2nd card is 2} \mid \text{1st card is 1}) = \tfrac{2}{3}$$

Now $P(\text{1st card is 1} \cap \text{2nd card is 2}) = \tfrac{1}{4} \times \tfrac{2}{3}$ and $P(\text{1st card is 1}) = \tfrac{1}{4}$

$$\therefore \quad \frac{P(\text{1st card is 1} \cap \text{2nd card is 2})}{P(\text{1st card is 1})} = P(\text{2nd card is 2} \mid \text{1st card is 1})$$

In general if A and B are two events then
$$P(A \cap B) = P(A) \times P(B \mid A)$$

Example

One coin is selected at random from two coins and tossed. One of the coins is biased so that a head is twice as likely as a tail and the other coin is fair. If the coin shows a head, what is the probability that it is the biased coin?

The two events we are interested in are the choice of coin and whether it shows a head.

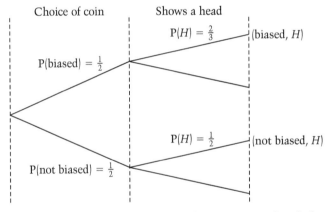

If B is the selection of the biased coin and H is tossing a head, then the probability that the coin is biased given that a head is tossed is $P(B \mid H)$.

Using $P(H \cap B) = P(H) \times P(B \mid H)$, then from the tree diagram

$$P(H \cap B) = \tfrac{1}{2} \times \tfrac{2}{3} = \tfrac{1}{3}$$

and
$$P(H) = \tfrac{1}{2} \times \tfrac{2}{3} + \tfrac{1}{2} \times \tfrac{1}{2} = \tfrac{7}{12}$$

$$\therefore \quad P(B \mid H) = \tfrac{1}{3} \div \tfrac{7}{12} = \tfrac{4}{7}$$

Exercise 3.8b

1 A bag contains 3 white balls and 2 black balls. A second bag contains 1 white ball and 4 black balls. One bag is chosen at random and one ball is removed at random from that bag.

   If the ball is black, what is the probability that it came out of the second bag?

2 A telephone call from one cell phone to another cell phone goes through three sets of independent equipment, the cell phone making the call, the operator's network and the cell phone receiving the call.

The probability that the outgoing phone is faulty is 0.001, the probability that the network is faulty is 0.01 and the probability that the receiving phone is faulty is 0.005

Find the probability that if a call fails to connect, it is at least partly a fault of the network.

3.9 Introduction to matrices

You need to know

- The meaning of commutative operations

Matrices

A *matrix* is an array of elements in rows and columns (numbers or algebraic expressions) that are enclosed in brackets, for example

$$\begin{pmatrix} 2 & -1 & 0 & 3 & -4 \\ -2 & 6 & 4 & 5 & 0 \end{pmatrix}$$

A matrix is denoted by **A**, **B**, etc.

The size of a matrix is defined by the number of rows and the number of columns, in that order. For example,

$$\begin{pmatrix} 2 & -1 & 0 & 3 & -4 \\ -2 & 6 & 4 & 5 & 0 \end{pmatrix}$$ has 2 rows and 5 columns and is called a 2×5 matrix.

When a matrix has m rows and n columns it is called an $m \times n$ matrix.

The position of a particular element is identified by suffixes to show which row and column (in that order) it is in. For example, a_{21} means the element in the second row and first column and a_{ij} means the element in the ith row and jth column. In the example above, $a_{21} = -2$

A matrix with just one column is called a **column vector** and a matrix with just one row is called a **row vector**. Column vectors and row vectors are denoted by **a**, **b**, etc.

For example, $\quad \mathbf{a} = \begin{pmatrix} 6 \\ 2 \\ -1 \end{pmatrix}$ and $\mathbf{b} = (-8 \ 5 \ 10)$

A matrix with the same number of rows and columns is called a **square matrix**, e.g. $\begin{pmatrix} 3 & 8 \\ 0 & -9 \end{pmatrix}$

Two matrices **A** and **B** are equal when each element in **A** is equal to the corresponding element in **B**, i.e.

$$\mathbf{A} = \mathbf{B} \iff a_{ij} = b_{ij} \text{ for all values of } i \text{ and } j$$

For example, $\begin{pmatrix} 3 & 8 \\ 0 & -9 \end{pmatrix} = \begin{pmatrix} 3 & 8 \\ 0 & -9 \end{pmatrix}$ but $\begin{pmatrix} 3 & 8 \\ 0 & -9 \end{pmatrix} \neq \begin{pmatrix} 3 & 6 \\ 1 & -9 \end{pmatrix}$

Example

Find x and y given that $\begin{pmatrix} x & 0 \\ -1 & 4 \end{pmatrix} = \begin{pmatrix} 3 & 0 \\ -1 & y \end{pmatrix}$

As $a_{ij} = b_{ij}$ for all values of i and j, $x = 3$ and $y = 4$

Addition and subtraction of matrices

Matrices can be added when they are the *same size* by adding corresponding elements.

For example, $\begin{pmatrix} 2 & -3 \\ 4 & 0 \\ 6 & -1 \end{pmatrix} + \begin{pmatrix} 4 & 0 \\ -2 & 7 \\ 5 & -5 \end{pmatrix} = \begin{pmatrix} 6 & -3 \\ 2 & 7 \\ 11 & -6 \end{pmatrix}$

The matrices $\begin{pmatrix} 2 & -1 \\ 8 & 2 \end{pmatrix}$ and $\begin{pmatrix} 4 & 1 \\ -2 & 6 \\ 4 & -1 \end{pmatrix}$ are different sizes and cannot be added, so their sum has no meaning.

The addition of real numbers is commutative, so if **A** and **B** are two matrices of the same size, it follows that

$$\mathbf{A} + \mathbf{B} = \mathbf{B} + \mathbf{A}$$

Matrices can be subtracted when they are the same size, by subtracting corresponding elements.

For example, $\begin{pmatrix} 2 & -3 \\ 4 & 0 \\ 6 & -1 \end{pmatrix} - \begin{pmatrix} 4 & 0 \\ -2 & 7 \\ 5 & -5 \end{pmatrix} = \begin{pmatrix} -2 & -3 \\ 6 & -7 \\ 1 & 4 \end{pmatrix}$

Subtraction of real numbers is not commutative, so if **A** and **B** are two matrices of the same size, it follows that

$$\mathbf{A} - \mathbf{B} \neq \mathbf{B} - \mathbf{A}$$

Matrices that are the same size are said to be ***conformable for addition and subtraction***.

Multiplication of a matrix by a scalar

The elements in the matrix $\lambda\mathbf{A}$ are each λ times the corresponding elements in **A**.

For example, when $\mathbf{A} = \begin{pmatrix} 2 & -3 \\ 4 & 0 \\ 6 & -1 \end{pmatrix}$, $3\mathbf{A} = \begin{pmatrix} 6 & -9 \\ 12 & 0 \\ 18 & -3 \end{pmatrix}$

Example

Given $\mathbf{A} = \begin{pmatrix} 5 & -4 & -1 \\ 2 & 5 & 0 \end{pmatrix}$ and $\mathbf{B} = \begin{pmatrix} -2 & 0 & -2 \\ 4 & -6 & -1 \end{pmatrix}$, find

(a) $\mathbf{A} + \mathbf{B}$

(b) $2\mathbf{A} - \mathbf{B}$

(a) $\mathbf{A} + \mathbf{B} = \begin{pmatrix} 5 & -4 & -1 \\ 2 & 5 & 0 \end{pmatrix} + \begin{pmatrix} -2 & 0 & -2 \\ 4 & -6 & -1 \end{pmatrix}$

$\qquad = \begin{pmatrix} 3 & -4 & -3 \\ 6 & -1 & -1 \end{pmatrix}$

(b) $\qquad 2\mathbf{A} = \begin{pmatrix} 10 & -8 & -2 \\ 4 & 10 & 0 \end{pmatrix}$, therefore

$2\mathbf{A} - \mathbf{B} = \begin{pmatrix} 10 & -8 & -2 \\ 4 & 10 & 0 \end{pmatrix} - \begin{pmatrix} -2 & 0 & -2 \\ 4 & -6 & -1 \end{pmatrix}$

$\qquad = \begin{pmatrix} 12 & -8 & 0 \\ 0 & 16 & 1 \end{pmatrix}$

Exercise 3.9

$\mathbf{A} = \begin{pmatrix} 4 & 5 \\ -2 & 0 \\ 3 & -6 \end{pmatrix}$, $\mathbf{B} = \begin{pmatrix} -1 & 2 \\ -3 & 2 \\ 3 & -1 \end{pmatrix}$, $\mathbf{C} = \begin{pmatrix} x & 2x & -x \\ y & xy & x^2 \\ 3x & -2x & xy \end{pmatrix}$, $\mathbf{D} = \begin{pmatrix} 5x & 2y & -x \\ y & xy & y^2 \\ 3y & -2x & 2y^2 \end{pmatrix}$

1 Use the given matrices to find:

   (a) $\mathbf{B} - \mathbf{A}$ (b) $3\mathbf{A} + 2\mathbf{B}$ (c) $\mathbf{D} - \mathbf{C}$

2 If c_{ij} is an element in **C**, write down the element c_{31}.

3 Explain why $\mathbf{A} + \mathbf{C}$ has no meaning.

Learning outcomes

- To define the product of a row vector and a column vector
- To define the product of a square matrix and a column vector
- To define the product of two matrices

You need to know

- The meanings of column matrix, row matrix, $m \times n$ matrices
- The notation for the elements of a matrix
- The double angle trig identities

The product of a row vector and a column vector

Provided that a row vector and a column vector each have the same number of elements,

the product $(a_{11} \ a_{12} \ \dots \ a_{1n})\begin{pmatrix} b_{11} \\ b_{21} \\ \vdots \\ b_{n1} \end{pmatrix}$ is defined as $a_{11}b_{11} + a_{12}b_{21} \ \dots \ + a_{1n}b_{n1}$

For example, $(2 \ 1 \ -1)\begin{pmatrix} 3 \\ 4 \\ -2 \end{pmatrix} = (2)(3) + (1)(4) + (-1)(-2) = 12$

The product of a matrix and a column vector

Provided that a matrix has the same number of columns as the number

of elements in a column vector, the product $\begin{pmatrix} a_{11} & \dots & a_{1n} \\ a_{21} & \dots & a_{2n} \\ \dots & \dots & \dots \\ a_{m1} & \dots & a_{mn} \end{pmatrix}\begin{pmatrix} b_{11} \\ \dots \\ b_{n1} \end{pmatrix}$

is defined as the column vector whose top element is the product of the top row of the matrix and the column vector, whose second element is the product of the second row of the matrix and the column vector, and so on,

i.e. $\begin{pmatrix} a_{11} & \dots & a_{1n} \\ a_{21} & \dots & a_{2n} \\ \dots & \dots & \dots \\ a_{m1} & \dots & a_{mn} \end{pmatrix}\begin{pmatrix} b_{11} \\ \dots \\ b_{n1} \end{pmatrix} = \begin{pmatrix} a_{11}b_{11} + a_{12}b_{21} + \dots + a_{1n}b_{n1} \\ a_{21}b_{11} + a_{22}b_{21} + \dots + a_{2n}b_{n1} \\ \dots \\ a_{m1}b_{11} + a_{m2}b_{21} + \dots + a_{mn}b_{n1} \end{pmatrix}$

For example,

$\begin{pmatrix} 5 & 2 & 7 \\ -1 & -6 & 2 \end{pmatrix}\begin{pmatrix} -1 \\ 2 \\ -5 \end{pmatrix} = \begin{pmatrix} (5 \ \ 2 \ \ 7)\begin{pmatrix} -1 \\ 2 \\ -5 \end{pmatrix} \\ (-1 \ -6 \ \ 2)\begin{pmatrix} -1 \\ 2 \\ -5 \end{pmatrix} \end{pmatrix}$

$= \begin{pmatrix} (5)(-1) + (2)(2) + (7)(-5) \\ (-1)(-1) + (-6)(2) + (2)(-5) \end{pmatrix}$

$= \begin{pmatrix} -36 \\ -21 \end{pmatrix}$

This definition of the product is precise: **Ab** exists when **b** is a column vector only if the number of columns in **A** is equal to the number of elements in **b**.

bA is meaningless and **bA** is said to be ***non-conformable***.

For any product of an $(m \times n)$ matrix by an $(n \times 1)$ column vector, the result is an $(m \times 1)$ column vector.

Example

Evaluate $\begin{pmatrix} 2 & x \\ -1 & y \\ 0 & z \end{pmatrix}\begin{pmatrix} -1 \\ 2 \end{pmatrix}$

$$\begin{pmatrix} 2 & x \\ -1 & y \\ 0 & z \end{pmatrix}\begin{pmatrix} -1 \\ 2 \end{pmatrix} = \begin{pmatrix} -2 + 2x \\ 1 + 2y \\ 2z \end{pmatrix}$$

Exercise 3.10a

Evaluate these products.

1 $(5 \quad -2)\begin{pmatrix} -4 \\ 2 \end{pmatrix}$ **2** $(2 \quad -5 \quad 1)\begin{pmatrix} 4 \\ 0 \\ -3 \end{pmatrix}$ **3** $(3 \quad -1 \quad 4)\begin{pmatrix} x \\ y \\ z \end{pmatrix}$

4 $\begin{pmatrix} 2 & 1 \\ 3 & -1 \end{pmatrix}\begin{pmatrix} -4 \\ 2 \end{pmatrix}$ **5** $\begin{pmatrix} 5 & 1 \\ -2 & -1 \\ 0 & 3 \end{pmatrix}\begin{pmatrix} 3 \\ -2 \end{pmatrix}$ **6** $\begin{pmatrix} 3 & -1 & 4 \\ 2 & -2 & 0 \\ 1 & 5 & -1 \end{pmatrix}\begin{pmatrix} x \\ y \\ z \end{pmatrix}$

The product of two matrices

For two matrices **A** and **B**, the product **AB** exists provided that the number of columns in **A** is equal to the number of rows in **B**. Two matrices that satisfy this condition are called **conformable for multiplication**.

AB is then defined as the matrix **C** where the element c_{ij} is the product of the ith row of **A** and the jth column of **B**, i.e.

$$\begin{pmatrix} a_{11} & \cdots & a_{1n} \\ a_{21} & \cdots & a_{2n} \\ \cdots & \cdots & \cdots \\ a_{m1} & \cdots & a_{mn} \end{pmatrix}\begin{pmatrix} b_{11} & \cdots & b_{1m} \\ b_{21} & \cdots & b_{2m} \\ \cdots & \cdots & \cdots \\ b_{n1} & \cdots & b_{nm} \end{pmatrix} = \begin{pmatrix} c_{11} & \cdots & c_{1n} \\ \cdots & c_{ij} & \cdots \\ c_{m1} & \cdots & c_{nm} \end{pmatrix}$$

where c_{ij} is the product of the ith row and jth column, i.e. $c_{ij} = (a_{i1} \cdots a_{in})\begin{pmatrix} b_{1j} \\ \vdots \\ b_{nj} \end{pmatrix}$

For example,

$$\begin{pmatrix} 3 & 1 \\ 1 & -1 \end{pmatrix}\begin{pmatrix} 1 & 3 & 2 \\ -1 & 1 & 2 \end{pmatrix} = \begin{pmatrix} (3 \quad 1)\begin{pmatrix} 1 \\ -1 \end{pmatrix} & (3 \quad 1)\begin{pmatrix} 3 \\ 1 \end{pmatrix} & (3 \quad 1)\begin{pmatrix} 2 \\ 2 \end{pmatrix} \\ (1 \quad -1)\begin{pmatrix} 1 \\ -1 \end{pmatrix} & (1 \quad -1)\begin{pmatrix} 3 \\ 1 \end{pmatrix} & (1 \quad -1)\begin{pmatrix} 2 \\ 2 \end{pmatrix} \end{pmatrix}$$

$$= \begin{pmatrix} 2 & 10 & 8 \\ 2 & 2 & 0 \end{pmatrix}$$

However, $\begin{pmatrix} 1 & 3 & 2 \\ -1 & 1 & 2 \end{pmatrix}\begin{pmatrix} 3 & 1 \\ 1 & -1 \end{pmatrix}$ is meaningless because the first matrix

has three columns and the second matrix has only two rows, i.e. the matrices are not conformable.

Now consider the matrices **A** and **B** where $\mathbf{A} = \begin{pmatrix} 2 & 4 & 1 \\ 1 & 0 & -1 \end{pmatrix}$ and $\mathbf{B} = \begin{pmatrix} 2 & 1 \\ 0 & -2 \\ -1 & 1 \end{pmatrix}$

The product **AB** exists because **A** has three columns and **B** has three rows, so

$$\mathbf{AB} = \begin{pmatrix} 2 & 4 & 1 \\ 1 & 0 & -1 \end{pmatrix} \begin{pmatrix} 2 & 1 \\ 0 & -2 \\ -1 & 1 \end{pmatrix}$$

$$= \begin{pmatrix} ((2)(2) + (4)(0) + (1)(-1)) & ((2)(1) + (4)(-2) + (1)(1)) \\ ((1)(2) + (0)(0) + (-1)(-1)) & ((1)(1) + (0)(-2) + (-1)(1)) \end{pmatrix}$$

$$= \begin{pmatrix} 3 & -5 \\ 3 & 0 \end{pmatrix}$$

The product **BA** also exists because **B** has two columns and **A** has two rows, so

$$\mathbf{BA} = \begin{pmatrix} 2 & 1 \\ 0 & -2 \\ -1 & 1 \end{pmatrix} \begin{pmatrix} 2 & 4 & 1 \\ 1 & 0 & -1 \end{pmatrix}$$

$$= \begin{pmatrix} (2)(2) + (1)(1) & (2)(4) + (1)(0) & (2)(1) + (1)(-1) \\ (0)(2) + (-2)(1) & (0)(4) + (-2)(0) & (0)(1) + (-2)(-1) \\ (-1)(2) + (1)(1) & (-1)(4) + (1)(0) & (-1)(1) + (1)(-1) \end{pmatrix}$$

$$= \begin{pmatrix} 5 & 8 & 1 \\ -2 & 0 & 2 \\ -1 & -4 & -2 \end{pmatrix}$$

This example illustrates the following key points.

> **When AB and BA both exist, in general AB ≠ BA so matrix multiplication is not commutative.**

> **The order in which the matrices are multiplied matters, so for AB we say that A premultiplies B and for BA we say that A postmultiplies B.**

> **When A is an $m \times n$ matrix and B is an $n \times p$ matrix, the size of AB is $m \times p$**

Note that $\mathbf{A}^2$ means $\mathbf{AA}$.

Example

Given $\mathbf{A} = \begin{pmatrix} 4 & -1 \\ 3 & 6 \end{pmatrix}$ and $\mathbf{I} = \begin{pmatrix} 1 & 0 \\ 0 & 1 \end{pmatrix}$ show that

(a) $\mathbf{AI} = \mathbf{IA}$ 　　　　　　　　 **(b)** $\mathbf{A}^2 - \mathbf{A} - 9\mathbf{I} = \begin{pmatrix} 0 & -9 \\ 27 & 18 \end{pmatrix}$

(a) $\mathbf{AI} = \begin{pmatrix} 4 & -1 \\ 3 & 6 \end{pmatrix} \begin{pmatrix} 1 & 0 \\ 0 & 1 \end{pmatrix} = \begin{pmatrix} 4 & -1 \\ 3 & 6 \end{pmatrix}$

and $\mathbf{IA} = \begin{pmatrix} 1 & 0 \\ 0 & 1 \end{pmatrix} \begin{pmatrix} 4 & -1 \\ 3 & 6 \end{pmatrix} = \begin{pmatrix} 4 & -1 \\ 3 & 6 \end{pmatrix}$

∴ $\mathbf{AI} = \mathbf{IA} \, (= \mathbf{A})$

(b) $\mathbf{A}^2 = \begin{pmatrix} 4 & -1 \\ 3 & 6 \end{pmatrix}\begin{pmatrix} 4 & -1 \\ 3 & 6 \end{pmatrix} = \begin{pmatrix} 13 & -10 \\ 30 & 33 \end{pmatrix}$

$\mathbf{A}^2 - \mathbf{A} = \begin{pmatrix} 13 & -10 \\ 30 & 33 \end{pmatrix} - \begin{pmatrix} 4 & -1 \\ 3 & 6 \end{pmatrix} = \begin{pmatrix} 9 & -9 \\ 27 & 27 \end{pmatrix}$

$\therefore \quad \mathbf{A}^2 - \mathbf{A} - 9\mathbf{I} = \begin{pmatrix} 9 & -9 \\ 27 & 27 \end{pmatrix} - \begin{pmatrix} 9 & 0 \\ 0 & 9 \end{pmatrix} = \begin{pmatrix} 0 & -9 \\ 27 & 18 \end{pmatrix}$

$\therefore \quad \mathbf{A}^2 - \mathbf{A} - 9\mathbf{I} = \begin{pmatrix} 0 & -9 \\ 27 & 18 \end{pmatrix}$

Example

Given $\mathbf{A} = \begin{pmatrix} 2 & -1 \\ 3 & 2 \end{pmatrix}$, $\mathbf{B} = \begin{pmatrix} 1 & 4 \\ 0 & 1 \end{pmatrix}$ and $\mathbf{C} = \begin{pmatrix} -2 & 0 \\ -2 & 3 \end{pmatrix}$

show that $(\mathbf{AB})\mathbf{C} = \mathbf{A}(\mathbf{BC})$

$\mathbf{AB} = \begin{pmatrix} 2 & -1 \\ 3 & 2 \end{pmatrix}\begin{pmatrix} 1 & 4 \\ 0 & 1 \end{pmatrix} = \begin{pmatrix} 2 & 7 \\ 3 & 14 \end{pmatrix}$,

$\therefore \quad (\mathbf{AB})\mathbf{C} = \begin{pmatrix} 2 & 7 \\ 3 & 14 \end{pmatrix}\begin{pmatrix} -2 & 0 \\ -2 & 3 \end{pmatrix} = \begin{pmatrix} -18 & 21 \\ -34 & 42 \end{pmatrix}$

$\mathbf{BC} = \begin{pmatrix} 1 & 4 \\ 0 & 1 \end{pmatrix}\begin{pmatrix} -2 & 0 \\ -2 & 3 \end{pmatrix} = \begin{pmatrix} -10 & 12 \\ -2 & 3 \end{pmatrix}$

$\therefore \quad \mathbf{A}(\mathbf{BC}) = \begin{pmatrix} 2 & -1 \\ 3 & 2 \end{pmatrix}\begin{pmatrix} -10 & 12 \\ -2 & 3 \end{pmatrix} = \begin{pmatrix} -18 & 21 \\ -34 & 42 \end{pmatrix}$

Therefore $(\mathbf{AB})\mathbf{C} = \mathbf{A}(\mathbf{BC})$

In general if the products can be found, then for three matrices A, B and C

$$(\mathbf{AB})\mathbf{C} = \mathbf{A}(\mathbf{BC})$$

i.e. matrix multiplication is associative.

Exercise 3.10b

1 Evaluate

(a) $\begin{pmatrix} -1 & 0 \\ 2 & 4 \end{pmatrix}\begin{pmatrix} 2 & 3 \\ -1 & 0 \end{pmatrix}$

(b) $\begin{pmatrix} 2 & 4 & 5 \\ -1 & 3 & 0 \\ 2 & -1 & 2 \end{pmatrix}\begin{pmatrix} 1 & 0 \\ -1 & 2 \\ 4 & 1 \end{pmatrix}$

(c) $\begin{pmatrix} -5 & 2 & 6 & -1 \\ 3 & -1 & 4 & -2 \end{pmatrix}\begin{pmatrix} 1 & 5 \\ 3 & 4 \\ -2 & -5 \\ 2 & 1 \end{pmatrix}$

2 Given $\mathbf{A} = \begin{pmatrix} 1 & 1 \\ 4 & -1 \end{pmatrix}$ and $\mathbf{B} = \begin{pmatrix} 1 & -1 \\ 2 & -1 \end{pmatrix}$ show that $(\mathbf{A} + \mathbf{B})^2 = \mathbf{A}^2 + \mathbf{B}^2$

3 Given $\mathbf{A} = \begin{pmatrix} \cos\dfrac{\pi}{2} & \sin\dfrac{\pi}{2} \\ \sin\dfrac{\pi}{2} & \cos\dfrac{\pi}{2} \end{pmatrix}$ show that $\mathbf{A}^2 = \begin{pmatrix} 1 & 0 \\ 0 & 1 \end{pmatrix}$

3.11 Square matrices, zero matrices, unit matrices and inverse matrices

Square matrices

A square matrix has an equal number of rows and columns.

For example $\begin{pmatrix} 2 & -4 \\ -1 & 3 \end{pmatrix}$ is a 2 × 2 square matrix and $\begin{pmatrix} -1 & 2 & 1 \\ 2 & 0 & 3 \\ 1 & -4 & -2 \end{pmatrix}$ is a 3 × 3 square matrix.

Unit matrices

A **unit matrix** is a square matrix such that the elements in the **leading diagonal** (that is top left to bottom right) are all 1 and all the other elements are zero.

For example, $\begin{pmatrix} 1 & 0 \\ 0 & 1 \end{pmatrix}$ and $\begin{pmatrix} 1 & 0 & 0 \\ 0 & 1 & 0 \\ 0 & 0 & 1 \end{pmatrix}$ are unit matrices.

A unit matrix is denoted by **I**.

Zero matrices

All the elements in a **zero matrix** are zero. A zero matrix is denoted by **0** and is not necessarily square, for example, $\begin{pmatrix} 0 & 0 & 0 \\ 0 & 0 & 0 \end{pmatrix}$ is a zero matrix.

Multiplication of a matrix by a zero matrix of a suitable size will give a zero matrix, for example, $\begin{pmatrix} a & c & e \\ b & d & f \end{pmatrix} \begin{pmatrix} 0 & 0 \\ 0 & 0 \\ 0 & 0 \end{pmatrix} = \begin{pmatrix} 0 & 0 \\ 0 & 0 \end{pmatrix}$

However, unlike real numbers where $ab = 0 \Rightarrow a = 0$ or $b = 0$, when $\mathbf{AB} = \mathbf{0}$, neither **A** nor **B** may be **0**.

For example, $\begin{pmatrix} 2 & -1 \\ 0 & 0 \end{pmatrix} \begin{pmatrix} 0 & 1 \\ 0 & 2 \end{pmatrix} = \begin{pmatrix} 0 & 0 \\ 0 & 0 \end{pmatrix}$

i.e. $\mathbf{AB} = \mathbf{0} \nRightarrow \mathbf{A} = \mathbf{0}$ or $\mathbf{B} = \mathbf{0}$

For real numbers, $ab = ac \Rightarrow a = 0$ or $b = c$

However, given $\mathbf{A} = \begin{pmatrix} 1 & 1 \\ 1 & 1 \end{pmatrix}$, $\mathbf{B} = \begin{pmatrix} 1 & 0 \\ 0 & 1 \end{pmatrix}$, $\mathbf{C} = \begin{pmatrix} 0 & 1 \\ 1 & 0 \end{pmatrix}$

$\mathbf{AB} = \begin{pmatrix} 1 & 1 \\ 1 & 1 \end{pmatrix} \begin{pmatrix} 1 & 0 \\ 0 & 1 \end{pmatrix} = \begin{pmatrix} 1 & 1 \\ 1 & 1 \end{pmatrix}$

and $\mathbf{AC} = \begin{pmatrix} 1 & 1 \\ 1 & 1 \end{pmatrix} \begin{pmatrix} 0 & 1 \\ 1 & 0 \end{pmatrix} = \begin{pmatrix} 1 & 1 \\ 1 & 1 \end{pmatrix}$

so $\mathbf{AB} = \mathbf{AC}$ but $\mathbf{A} \neq \mathbf{0}$ and $\mathbf{B} \neq \mathbf{C}$

i.e. $\mathbf{AB} = \mathbf{AC} \nRightarrow \mathbf{A} = \mathbf{0}$ or $\mathbf{B} = \mathbf{C}$

Identity matrices

Under addition and subtraction of real numbers, the identity number is 0 as it leaves unchanged any number it is added to or subtracted from.

There is no single *identity matrix* that can be added to or subtracted from any matrix that will leave that matrix unchanged.

The identity matrix for addition and subtraction is a zero matrix of the same size.

For example,

$$\begin{pmatrix} a & c & e \\ b & d & f \end{pmatrix} \pm \begin{pmatrix} 0 & 0 & 0 \\ 0 & 0 & 0 \end{pmatrix} = \begin{pmatrix} a & c & e \\ b & d & f \end{pmatrix} \text{ but } \begin{pmatrix} a & c & e \\ b & d & f \end{pmatrix} \pm \begin{pmatrix} 0 & 0 \\ 0 & 0 \end{pmatrix}$$

is meaningless.

Under multiplication of real numbers, the identity number is 1 as it leaves unchanged any number multiplied by 1.

Again, there is no single identity matrix that can premultiply and/or postmultiply any matrix and leave it unchanged.

The identity matrix for multiplication is a unit matrix and its size depends on the matrix it multiplies.

For a square matrix, the identity is the same size. For example,

$$\begin{pmatrix} a & c \\ b & d \end{pmatrix}\begin{pmatrix} 1 & 0 \\ 0 & 1 \end{pmatrix} = \begin{pmatrix} a & c \\ b & d \end{pmatrix} \text{ and } \begin{pmatrix} 1 & 0 \\ 0 & 1 \end{pmatrix}\begin{pmatrix} a & c \\ b & d \end{pmatrix} = \begin{pmatrix} a & c \\ b & d \end{pmatrix}$$

$$\begin{pmatrix} a & d & g \\ b & e & h \\ c & f & i \end{pmatrix}\begin{pmatrix} 1 & 0 & 0 \\ 0 & 1 & 0 \\ 0 & 0 & 1 \end{pmatrix} = \begin{pmatrix} a & d & g \\ b & e & h \\ c & f & i \end{pmatrix} \text{ and } \begin{pmatrix} 1 & 0 & 0 \\ 0 & 1 & 0 \\ 0 & 0 & 1 \end{pmatrix}\begin{pmatrix} a & d & g \\ b & e & h \\ c & f & i \end{pmatrix} = \begin{pmatrix} a & d & g \\ b & e & h \\ c & f & i \end{pmatrix}$$

Therefore, a unit matrix of the same size as a square matrix is both a premultiplicative and a postmultiplicative identity.

Now consider the matrix $\mathbf{A} = \begin{pmatrix} a & c & e \\ b & d & f \end{pmatrix}$

We can premultiply $\mathbf{A}$ by the unit matrix $\begin{pmatrix} 1 & 0 \\ 0 & 1 \end{pmatrix}$ giving

$$\begin{pmatrix} 1 & 0 \\ 0 & 1 \end{pmatrix}\begin{pmatrix} a & c & e \\ b & d & f \end{pmatrix} = \begin{pmatrix} a & c & e \\ b & d & f \end{pmatrix}$$

and we can postmultiply $\mathbf{A}$ by the unit matrix $\begin{pmatrix} 1 & 0 & 0 \\ 0 & 1 & 0 \\ 0 & 0 & 1 \end{pmatrix}$ giving

$$\begin{pmatrix} a & c & e \\ b & d & f \end{pmatrix}\begin{pmatrix} 1 & 0 & 0 \\ 0 & 1 & 0 \\ 0 & 0 & 1 \end{pmatrix} = \begin{pmatrix} a & c & e \\ b & d & f \end{pmatrix}$$

In general, for an $m \times n$ matrix, an $m \times m$ unit matrix is a premultiplicative identity and an $n \times n$ unit matrix is a postmultiplicative identity.

Exercise 3.11

1 Given $\mathbf{A} = \begin{pmatrix} i & 0 \\ 0 & i \end{pmatrix}$ where $i = \sqrt{-1}$ show that $\mathbf{A}^4 = \mathbf{I}$

Learning outcomes

- To define determinants
- To calculate the determinant of an $n \times n$ matrix
- To define a singular matrix

You need to know

- The notation for identifying elements in a matrix
- The meaning of the leading diagonal in a square matrix

Determinant of a matrix

The **determinant** of a square matrix is a real number that is associated with that matrix. Only square matrices have determinants.

The determinant of a matrix $\mathbf{A} = \begin{pmatrix} a_{11} & \dots & \dots & a_{1n} \\ a_{21} & \dots & \dots & a_{2n} \\ \dots & \dots & \dots & \dots \\ a_{n1} & \dots & \dots & a_{nn} \end{pmatrix}$ is denoted by

$$|\mathbf{A}| = \begin{vmatrix} a_{11} & \dots & \dots & a_{1n} \\ a_{21} & \dots & \dots & a_{2n} \\ \dots & \dots & \dots & \dots \\ a_{n1} & \dots & \dots & a_{nn} \end{vmatrix}$$

$|\mathbf{A}|$ is also written as $\det \mathbf{A}$.

The determinant of a 2 × 2 matrix

The determinant of the matrix $\begin{pmatrix} a & b \\ c & d \end{pmatrix}$ is defined as the value of $ad - bc$, i.e. as the value of the product of the elements in the leading diagonal minus the product of the elements in the other diagonal.

$$\textbf{Therefore} \quad \begin{vmatrix} a & b \\ c & d \end{vmatrix} = ad - bc$$

For example, if $\mathbf{A} = \begin{pmatrix} 2 & -1 \\ 3 & 5 \end{pmatrix}$ then $|\mathbf{A}| = \begin{vmatrix} 2 & -1 \\ 3 & 5 \end{vmatrix} = (2)(5) - (-1)(3) = 13$

Exercise 3.12a

Find the determinant of each matrix.

1   $\begin{pmatrix} 3 & 2 \\ -1 & -2 \end{pmatrix}$ 2 $\begin{pmatrix} 6 & -2 \\ -5 & 1 \end{pmatrix}$ 3 $\begin{pmatrix} \frac{1}{2} & \frac{2}{3} \\ \frac{5}{6} & \frac{10}{9} \end{pmatrix}$ 4 $\begin{pmatrix} 2x & y \\ x & x^2 \end{pmatrix}$

Cofactors

The determinant of an $n \times n$ matrix is based on extracting smaller determinants.

The determinant of a 3×3 matrix is based on 2×2 determinants extracted from the 3×3 determinant.

These 2×2 determinants are found from the elements left when the row and column through a particular entry are crossed out.

For example, when we cross out the elements from the row and column through the element 8 in $\begin{vmatrix} 2 & 1 & 3 \\ 6 & 4 & 8 \\ 5 & 9 & 7 \end{vmatrix}$

we are left with the 2×2 determinant $\begin{vmatrix} 2 & 1 \\ 5 & 9 \end{vmatrix}$

This determinant is called the ***minor*** of the element 8.

Each minor in a determinant has a sign, $+$ or $-$, associated with it. This sign depends on the position of the element of which it is minor.

These signs are $\begin{vmatrix} + & - & + \\ - & + & - \\ + & - & + \end{vmatrix}$

The minor of an element together with its sign is called the
***cofactor* of that element.**

So, for example, the sign $-$ is associated with the element 8 in $\begin{vmatrix} 2 & 1 & 3 \\ 6 & 4 & 8 \\ 5 & 9 & 7 \end{vmatrix}$

so the cofactor of 8 is $-\begin{vmatrix} 2 & 1 \\ 5 & 9 \end{vmatrix}$

The cofactors of 2, 1 and 3 are $\begin{vmatrix} 4 & 8 \\ 9 & 7 \end{vmatrix}$, $-\begin{vmatrix} 6 & 8 \\ 5 & 7 \end{vmatrix}$ and $\begin{vmatrix} 6 & 4 \\ 5 & 9 \end{vmatrix}$ respectively.

The determinant of a 3 × 3 matrix

The determinant of a 3×3 matrix is defined as the sum of the products of each element in the first row and its cofactor.

So, for example, $\begin{vmatrix} 2 & 1 & 3 \\ 6 & 4 & 8 \\ 5 & 9 & 7 \end{vmatrix} = 2\begin{vmatrix} 4 & 8 \\ 9 & 7 \end{vmatrix} - 1\begin{vmatrix} 6 & 8 \\ 5 & 7 \end{vmatrix} + 3\begin{vmatrix} 6 & 4 \\ 5 & 9 \end{vmatrix}$

$$= (2)(-44) - (1)(2) + (3)(34) = 12$$

In general

$$\begin{vmatrix} a_{11} & a_{12} & a_{13} \\ a_{21} & a_{22} & a_{23} \\ a_{31} & a_{32} & a_{33} \end{vmatrix} = a_{11}\begin{vmatrix} a_{22} & a_{23} \\ a_{32} & a_{33} \end{vmatrix} - a_{12}\begin{vmatrix} a_{21} & a_{23} \\ a_{31} & a_{33} \end{vmatrix} + a_{13}\begin{vmatrix} a_{21} & a_{22} \\ a_{31} & a_{32} \end{vmatrix}$$

It is not sensible to try and remember this as a general method, just remember the process.

The determinants of larger matrices are found by extending the definition of a 3×3 determinant, i.e. by the sum of the products of the elements in the first row and their cofactors.

For a 4×4 determinant, the cofactors are 3×3 determinants, and these in turn are broken down to 2×2 determinants. However, the calculation involved in finding a determinant larger than 3×3 is tedious and is usually done using a specialist calculator or software.

 Exam tip

It is easy to make mistakes when finding a 3×3 determinant, so do not be tempted to try and calculate it in one step.

Example

Evaluate $\begin{vmatrix} -1 & 2 & 0 \\ -3 & 1 & -1 \\ 2 & 0 & 5 \end{vmatrix}$

$$\begin{vmatrix} -1 & 2 & 0 \\ -3 & 1 & -1 \\ 2 & 0 & 5 \end{vmatrix} = (-1)\begin{vmatrix} 1 & -1 \\ 0 & 5 \end{vmatrix} - (2)\begin{vmatrix} -3 & -1 \\ 2 & 5 \end{vmatrix} + (0)\begin{vmatrix} -3 & 1 \\ 2 & 0 \end{vmatrix}$$

$$= (-1)(5) - (2)(-13)$$

$$= 21$$

Example

Solve the equation $\begin{vmatrix} 2 & x & 1 \\ x & 3 & -1 \\ 4 & 1 & -1 \end{vmatrix} = 24$

Expanding the determinant gives

$$\begin{vmatrix} 2 & x & 1 \\ x & 3 & -1 \\ 4 & 1 & -1 \end{vmatrix} = 2(-3 + 1) - x(-x + 4) + 1(x - 12)$$

$$= -4 + x^2 - 4x + x - 12$$

$$= x^2 - 3x - 16$$

$\therefore \quad x^2 - 3x - 16 = 24$

$\Rightarrow \quad x^2 - 3x - 40 = 0$

$\Rightarrow \quad (x - 8)(x + 5) = 0$

$\therefore \quad x = -5 \text{ or } x = 8$

An application of 3 × 3 determinants in coordinate geometry

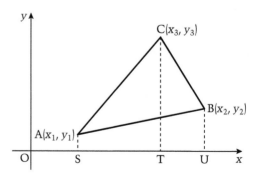

The area of the triangle ABC in the diagram is
(area SACT) + (area TCBU) − (area SABU)

$$= \tfrac{1}{2}(y_1 + y_3)(x_3 - x_1) + \tfrac{1}{2}(y_2 + y_3)(x_2 - x_3) - \tfrac{1}{2}(y_1 + y_2)(x_2 - x_1)$$

$$= \tfrac{1}{2}(x_2 y_3 - x_3 y_2 - x_1 y_3 + x_3 y_1 + x_1 y_2 - x_2 y_1)$$

Writing this as

$$= \tfrac{1}{2}\left((1)(x_2 y_3 - x_3 y_2) - (1)(x_1 y_3 - x_3 y_1) + (1)(x_1 y_2 - x_2 y_1)\right)$$

shows that it is the expansion of the determinant

$$\frac{1}{2}\begin{vmatrix} 1 & 1 & 1 \\ x_1 & x_2 & x_3 \\ y_1 & y_2 & y_3 \end{vmatrix}$$

Hence the area of a triangle whose vertices are at the points (x_1, y_1), (x_2, y_2) and (x_3, y_3) is

$$\frac{1}{2}\begin{vmatrix} 1 & 1 & 1 \\ x_1 & x_2 & x_3 \\ y_1 & y_2 & y_3 \end{vmatrix}$$

Also, if the points A, B and C are collinear, the area of triangle ABC is zero, so

the condition for three points (x_1, y_1), (x_2, y_2) and (x_3, y_3) to be collinear is

$$\begin{vmatrix} 1 & 1 & 1 \\ x_1 & x_2 & x_3 \\ y_1 & y_2 & y_3 \end{vmatrix} = 0$$

Exercise 3.12b

1 Calculate each determinant.

(a) $\begin{vmatrix} 2 & 1 & 7 \\ 0 & -1 & 2 \\ 1 & 2 & 3 \end{vmatrix}$ (b) $\begin{vmatrix} 1 & 4 & -3 \\ 0 & 5 & 2 \\ 0 & 3 & -1 \end{vmatrix}$ (c) $\begin{vmatrix} -1 & 0 & 1 \\ -3 & 2 & 0 \\ 1 & -2 & 4 \end{vmatrix}$

2 Expand and simplify the determinant $\begin{vmatrix} 1 & 1 & 1 \\ \cos\theta & \cos^2\theta & 1 \\ \sin\theta & \sin^2\theta & 1 \end{vmatrix}$

3 Show that $\begin{vmatrix} a & b & c \\ a^2 & b^2 & c^2 \\ a^3 & b^3 & c^3 \end{vmatrix} = abc(a-b)(b-c)(c-a)$

4 Determine whether the following points are collinear.
(a) $(0, -6)$, $(1, -3)$, $(3, 3)$
(b) $(0, 1)$, $(1, 0)$, $(1, 1)$

- To simplify determinants

- How to find a 3×3 determinant
- The meaning of a cofactor

Simplification of determinants

When the elements in a determinant are large numbers or complicated algebraic expressions, it is easy to make mistakes when evaluating the determinant. However, there are properties of determinants that can be used to reduce elements to more manageable quantities.

Transposing the rows and columns of $|\mathbf{A}| = \begin{vmatrix} a_1 & a_2 & a_3 \\ b_1 & b_2 & b_3 \\ c_1 & c_2 & c_3 \end{vmatrix}$ gives $\begin{vmatrix} a_1 & b_1 & c_1 \\ a_2 & b_2 & c_2 \\ a_3 & b_3 & c_3 \end{vmatrix}$

which we denote by $|\mathbf{A^T}|$.

$$\text{Now} \quad |\mathbf{A}| = a_1(b_2c_3 - b_3c_2) - a_2(b_1c_3 - b_3c_1) + a_3(b_1c_2 - b_2c_1)$$
$$= a_1(b_2c_3 - b_3c_2) - b_1(a_2c_3 - a_3c_2) + c_1(a_2b_3 - a_3b_2) = |\mathbf{A^T}|$$

Therefore the value of a determinant is not changed when the rows and columns are transposed.
Hence any property proved for rows is also valid for columns.

The following properties can be proved using a method similar to that above.

A determinant can be expanded using any row or column and the respective cofactors.

For example, $\begin{vmatrix} 1 & 3 & -4 \\ 0 & -2 & 6 \\ 0 & 1 & 2 \end{vmatrix}$ can be found from the first column giving

$$(1)\begin{vmatrix} -2 & 6 \\ 1 & 2 \end{vmatrix} + 0 + 0 = -10$$

The value of a determinant is unchanged when any row (or column) is added to or subtracted from any other row (or column).

For example, $\begin{vmatrix} 2 & 3 & 5 \\ -1 & 2 & -1 \\ 2 & 3 & 2 \end{vmatrix} = \begin{vmatrix} 0 & 0 & 3 \\ -1 & 2 & -1 \\ 2 & 3 & 2 \end{vmatrix}$ Subtracting the third row from the first row

It follows from this property that if two rows or columns are the same, simplification will give a row or column of zeroes, so the determinant is equal to zero.

A matrix whose determinant is zero is called a **singular matrix**.

The value of a determinant is unchanged when a multiple of any row (or column) is added to or subtracted from any other row (or column).

For example, $\begin{vmatrix} 2 & 1 & -4 \\ -1 & 2 & 2 \\ 2 & 3 & -1 \end{vmatrix} = \begin{vmatrix} 2 & 1 & 0 \\ -1 & 2 & 0 \\ 2 & 3 & 3 \end{vmatrix}$ Adding twice the first column to the third column

The aim when simplifying a determinant is to get as many zeroes as possible in one row or column to make the evaluation easier and with less risk of mistakes, but be careful that you do not overdo it. It is also easy to make mistakes when adding and subtracting multiples of rows or columns.

Example

Solve the equation $\begin{vmatrix} 1 & 2 & 1 \\ x & x-1 & x+1 \\ 2x & 2x+1 & x-1 \end{vmatrix} = 0$

Subtracting the first column from the second and from the third column simplifies the elements containing x and gives a zero in the top row.

$$\begin{vmatrix} 1 & 2 & 1 \\ x & x-1 & x+1 \\ 2x & 2x+1 & x-1 \end{vmatrix} = \begin{vmatrix} 1 & 1 & 0 \\ x & -1 & 1 \\ 2x & 1 & -x-1 \end{vmatrix} = 0$$

Expanding the determinant gives

$(1)\begin{vmatrix} -1 & 1 \\ 1 & -x-1 \end{vmatrix} - 1\begin{vmatrix} x & 1 \\ 2x & -x-1 \end{vmatrix} + 0 = 0$

$\Rightarrow \qquad (x+1) - 1 - (-x^2 - x - 2x) = 0 \Rightarrow x^2 + 4x = 0$

$\Rightarrow \qquad\qquad\qquad x(x+4) = 0$

$\Rightarrow \qquad\qquad\qquad x = 0 \text{ or } x = -4$

Example

Given $f(x, y, z) = \begin{vmatrix} 1+x^2 & x & 1 \\ 1+y^2 & y & 1 \\ 1+z^2 & z & 1 \end{vmatrix}$ show that $(x - y)$ is a factor of the function f.

Subtracting the top row from the second and third row gives

$$f(x, y, z) = \begin{vmatrix} 1+x^2 & x & 1 \\ y^2 - x^2 & y-x & 0 \\ z^2 - x^2 & z-x & 0 \end{vmatrix}$$

Expanding using the third column and its cofactors gives

$f(x, y, z) = (1)((y^2 - x^2)(z - x) - (y - x)(z^2 - x^2))$

$\qquad\qquad = (y - x)((y + x)(z - x) - (z^2 - x^2))$

$\qquad\qquad = (x - y)((z^2 - x^2) - (y + x)(z - x))$

Therefore $(x - y)$ is a factor of the function f.

Exercise 3.13

1 Evaluate **(a)** $\begin{vmatrix} 1 & 6 & -10 \\ 2 & 8 & 16 \\ 1 & 8 & -14 \end{vmatrix}$ **(b)** $\begin{vmatrix} 100 & 200 & -100 \\ 20 & -18 & 16 \\ 21 & 36 & -14 \end{vmatrix}$

2 Express $\begin{vmatrix} x & 1 & x^2 \\ x^2 & 1 & x \\ x^3 & 1 & x^3 \end{vmatrix}$ as a product of factors.

Learning outcomes

- To find the multiplicative inverse of a matrix

You need to know

- How to multiply matrices
- What a unit matrix is
- How to evaluate a determinant
- The meaning of a singular matrix
- The effect of multiplying a matrix by a scalar
- How to find the cofactor of an element in a matrix

The meaning of a multiplicative inverse of a matrix

If for a matrix $\mathbf{A}$, a matrix $\mathbf{B}$ exists so that $\mathbf{AB} = \mathbf{I}$, $\mathbf{B}$ is called the *multiplicative inverse* of $\mathbf{A}$.

$\mathbf{B}$ is denoted by $\mathbf{A}^{-1}$, so $\mathbf{AA}^{-1} = \mathbf{I}$, and we will show that when $\mathbf{A}^{-1}$ exists, $\mathbf{AA}^{-1} = \mathbf{A}^{-1}\mathbf{A} = \mathbf{I}$.

This is similar to a multiplicative inverse for real numbers. Multiplying any real number by its reciprocal gives 1, e.g. $2 \times \frac{1}{2} = 1$, so $\frac{1}{2}$ is the multiplicative inverse of 2, and vice-versa.

In future we will call a multiplicative inverse of a matrix simply an *inverse matrix*.

The inverse of a 2 × 2 matrix

If $\mathbf{A} = \begin{pmatrix} a & b \\ c & d \end{pmatrix}$ then postmultiplying $\mathbf{A}$ by the matrix $\begin{pmatrix} d & -b \\ -c & a \end{pmatrix}$ gives

$$\begin{pmatrix} a & b \\ c & d \end{pmatrix}\begin{pmatrix} d & -b \\ -c & a \end{pmatrix} = \begin{pmatrix} ad - bc & 0 \\ 0 & ad - bc \end{pmatrix}$$

$$= (ad - bc) \times \begin{pmatrix} 1 & 0 \\ 0 & 1 \end{pmatrix}$$

Premultiplying $\mathbf{A}$ by $\begin{pmatrix} d & -b \\ -c & a \end{pmatrix}$ gives

$$\begin{pmatrix} d & -b \\ -c & a \end{pmatrix}\begin{pmatrix} a & b \\ c & d \end{pmatrix} = \begin{pmatrix} ad - bc & 0 \\ 0 & ad - bc \end{pmatrix}$$

$$= (ad - bc) \times \begin{pmatrix} 1 & 0 \\ 0 & 1 \end{pmatrix}$$

Now $ad - bc = |\mathbf{A}|$

Therefore both premultiplying and postmultiplying $\mathbf{A}$ by $\begin{pmatrix} d & -b \\ -c & a \end{pmatrix}$ gives

$|\mathbf{A}|\mathbf{I}$ so both premultiplying and postmultiplying $\mathbf{A}$ by $\frac{1}{|\mathbf{A}|}\begin{pmatrix} d & -b \\ -c & a \end{pmatrix}$ gives $\mathbf{I}$.

Therefore   when $A = \begin{pmatrix} a & b \\ c & d \end{pmatrix}$, $A^{-1} = \frac{1}{|A|}\begin{pmatrix} d & -b \\ -c & a \end{pmatrix}$

If $|A| = 0$, A^{-1} does not exist and A is a singular matrix.

Notice that the matrix $\begin{pmatrix} d & -b \\ -c & a \end{pmatrix}$ is obtained from the matrix $\begin{pmatrix} a & b \\ c & d \end{pmatrix}$ by transposing the elements in the leading diagonal and changing the signs of the elements in the other diagonal.

Example

Find the inverse of   **(a)** $\begin{pmatrix} 2 & 1 \\ 5 & 3 \end{pmatrix}$ **(b)** $\begin{pmatrix} 2 & 5 \\ 3 & 4 \end{pmatrix}$

(a) First transpose the elements in the leading diagonal and change
the sign of the other elements to give $\begin{pmatrix} 3 & -1 \\ -5 & 2 \end{pmatrix}$

Then $\begin{vmatrix} 2 & 1 \\ 5 & 3 \end{vmatrix} = 1$

therefore $\begin{pmatrix} 2 & 1 \\ 5 & 3 \end{pmatrix}^{-1} = \begin{pmatrix} 3 & -1 \\ -5 & 2 \end{pmatrix}$

(b) First transpose the elements in the leading diagonal and change
the sign of the other elements to give $\begin{pmatrix} 4 & -5 \\ -3 & 2 \end{pmatrix}$

Then $\begin{vmatrix} 2 & 5 \\ 3 & 4 \end{vmatrix} = -7$

therefore $\begin{pmatrix} 2 & 5 \\ 3 & 4 \end{pmatrix}^{-1} = -\dfrac{1}{7}\begin{pmatrix} 4 & -5 \\ -3 & 2 \end{pmatrix}$

$$= \begin{pmatrix} -\frac{4}{7} & \frac{5}{7} \\ \frac{3}{7} & -\frac{2}{7} \end{pmatrix}$$

Note that each of these answers can be checked by multiplying it
by the original matrix.

Exercise 3.14a

Find, when it exists, the inverse of each matrix.

1 $\begin{pmatrix} 4 & -2 \\ 1 & 1 \end{pmatrix}$ **2** $\begin{pmatrix} -4 & -2 \\ 4 & 1 \end{pmatrix}$ **3** $\begin{pmatrix} 2 & -3 \\ -4 & 6 \end{pmatrix}$ **4** $\begin{pmatrix} \sin\theta & \cos\theta \\ \cos\theta & -\sin\theta \end{pmatrix}$

The inverse of a 3 × 3 matrix

The inverse of the matrix $\mathbf{A} = \begin{pmatrix} a_1 & a_2 & a_3 \\ b_1 & b_2 & b_3 \\ c_1 & c_2 & c_3 \end{pmatrix}$ is found by first transposing

the rows and columns to give $\begin{pmatrix} a_1 & b_1 & c_1 \\ a_2 & b_2 & c_2 \\ a_3 & b_3 & c_3 \end{pmatrix}$ This is denoted by $\mathbf{A}^{\mathsf{T}}$

then replacing each element of **A** with its cofactor. Denoting the cofactor

of a_1 by A_1 and so on, this gives the matrix $\begin{pmatrix} A_1 & B_1 & C_1 \\ A_2 & B_2 & C_2 \\ A_3 & B_3 & C_3 \end{pmatrix}$ and finally

dividing by $|\mathbf{A}|$, i.e.

$$\text{when} \quad \mathbf{A} = \begin{pmatrix} a_1 & a_2 & a_3 \\ b_1 & b_2 & b_3 \\ c_1 & c_2 & c_3 \end{pmatrix}$$

$$\mathbf{A}^{-1} = \frac{1}{|\mathbf{A}|} \begin{pmatrix} A_1 & B_1 & C_1 \\ A_2 & B_2 & C_2 \\ A_3 & B_3 & C_3 \end{pmatrix}$$

where A_1, A_2, ... are the cofactors of a_1, a_2, ...

If $|\mathbf{A}| = 0$, $\mathbf{A}^{-1}$ does not exist and A is a singular matrix.

For example, to find the inverse of $\mathbf{A} = \begin{pmatrix} 4 & 1 & 0 \\ -1 & 2 & 1 \\ 3 & -2 & 1 \end{pmatrix}$, first find $\mathbf{A}^\mathrm{T}$.

$$\mathbf{A}^\mathrm{T} = \begin{pmatrix} 4 & -1 & 3 \\ 1 & 2 & -2 \\ 0 & 1 & 1 \end{pmatrix}$$

Next replace each element in $\mathbf{A}^\mathrm{T}$ by its cofactor:

$$\begin{pmatrix} +\begin{vmatrix} 2 & -2 \\ 1 & 1 \end{vmatrix} & -\begin{vmatrix} 1 & -2 \\ 0 & 1 \end{vmatrix} & +\begin{vmatrix} 1 & 2 \\ 0 & 1 \end{vmatrix} \\ -\begin{vmatrix} -1 & 3 \\ 1 & 1 \end{vmatrix} & +\begin{vmatrix} 4 & 3 \\ 0 & 1 \end{vmatrix} & -\begin{vmatrix} 4 & -1 \\ 0 & 1 \end{vmatrix} \\ +\begin{vmatrix} -1 & 3 \\ 2 & -2 \end{vmatrix} & -\begin{vmatrix} 4 & 3 \\ 1 & -2 \end{vmatrix} & +\begin{vmatrix} 4 & -1 \\ 1 & 2 \end{vmatrix} \end{pmatrix} = \begin{pmatrix} 4 & -1 & 1 \\ 4 & 4 & -4 \\ -4 & 11 & 9 \end{pmatrix}$$

Then find $|\mathbf{A}|$: $\quad \begin{vmatrix} 4 & 1 & 0 \\ -1 & 2 & 1 \\ 3 & -2 & 1 \end{vmatrix} = \begin{vmatrix} 4 & 1 & 0 \\ -4 & 4 & 0 \\ 3 & -2 & 1 \end{vmatrix}$

Subtracting the third row from the second row

$$= 20$$

Therefore $\quad \mathbf{A}^{-1} = \dfrac{1}{20} \begin{pmatrix} 4 & -1 & 1 \\ 4 & 4 & -4 \\ -4 & 11 & 9 \end{pmatrix}$

Check: $\begin{pmatrix} 4 & 1 & 0 \\ -1 & 2 & 1 \\ 3 & -2 & 1 \end{pmatrix} \times \dfrac{1}{20} \begin{pmatrix} 4 & -1 & 1 \\ 4 & 4 & -4 \\ -4 & 11 & 9 \end{pmatrix}$

$$= \frac{1}{20} \begin{pmatrix} 20 & 0 & 0 \\ 0 & 20 & 0 \\ 0 & 0 & 20 \end{pmatrix}$$

$$= \begin{pmatrix} 1 & 0 & 0 \\ 0 & 1 & 0 \\ 0 & 0 & 1 \end{pmatrix}$$

Note that it is very easy to make mistakes using this process, so take one step at a time.

Example

Show that the matrix $\mathbf{A} = \begin{pmatrix} 3 & 1 & 7 \\ 1 & 2 & 3 \\ 2 & -1 & 4 \end{pmatrix}$ does not have an inverse.

To find the inverse of any matrix $\mathbf{A}$, we need to evaluate $|\mathbf{A}|$. Therefore it is sensible to start with evaluating this determinant and if it is zero we know that the inverse does not exist.

$$|\mathbf{A}| = \begin{vmatrix} 3 & 1 & 7 \\ 1 & 2 & 3 \\ 2 & -1 & 4 \end{vmatrix}$$

Row 1 − row 2 gives $\begin{vmatrix} 2 & -1 & 4 \\ 1 & 2 & 3 \\ 2 & -1 & 4 \end{vmatrix}$

Rows 1 and 3 are the same, so taking one from the other will give a row of zeroes. Therefore $|\mathbf{A}| = 0$, hence $\mathbf{A}$ does not have an inverse.

Properties of inverse matrices

This is the property we proved for 2×2 matrices and it is true for all square matrices:

$$\mathbf{A}\mathbf{A}^{-1} = \mathbf{A}^{-1}\mathbf{A}$$

It follows from this that as $\mathbf{A}$ is the inverse of $\mathbf{A}^{-1}$, i.e.

$$(\mathbf{A}^{-1})^{-1} = \mathbf{A}$$

Now postmultiplying $\mathbf{AB}$ by $\mathbf{B}^{-1}\mathbf{A}^{-1}$ gives $(\mathbf{AB})(\mathbf{B}^{-1}\mathbf{A}^{-1})$

Matrix multiplication is associative, therefore

$$(\mathbf{AB})(\mathbf{B}^{-1}\mathbf{A}^{-1}) = \mathbf{A}(\mathbf{BB}^{-1})\mathbf{A}^{-1} = \mathbf{AIA}^{-1} = \mathbf{I}$$

Hence the inverse of $\mathbf{AB}$ is $\mathbf{B}^{-1}\mathbf{A}^{-1}$, i.e.

$$(\mathbf{AB})^{-1} = \mathbf{B}^{-1}\mathbf{A}^{-1}$$

Exercise 3.14b

1 Find, when it exists, the inverse of the matrices

(a) $\mathbf{A} = \begin{pmatrix} 1 & 1 & 0 \\ -1 & 3 & 0 \\ 2 & -1 & 1 \end{pmatrix}$ 　　(c) $\mathbf{B} = \begin{pmatrix} 1 & 0 & 4 \\ 0 & 3 & 2 \\ -1 & 5 & 0 \end{pmatrix}$

(b) $\begin{pmatrix} 5 & -1 & 9 \\ 3 & -1 & 5 \\ 2 & 0 & 4 \end{pmatrix}$ 　　(d) $\begin{pmatrix} 4 & 1 & 3 \\ 3 & 5 & -2 \\ -1 & -2 & 1 \end{pmatrix}$

2 Verify that, for the matrices $\mathbf{A}$ and $\mathbf{B}$ in question 1, $(\mathbf{AB})^{-1} = \mathbf{B}^{-1}\mathbf{A}^{-1}$

3 $\mathbf{A}$, $\mathbf{B}$ and $\mathbf{C}$ are non-singular 3×3 matrices.
Prove that $(\mathbf{ABC})^{-1} = \mathbf{C}^{-1}\mathbf{B}^{-1}\mathbf{A}^{-1}$

Learning outcomes

- To investigate the consistency of a pair of simultaneous equations in two unknowns
- To use matrices to solve a pair of simultaneous equations in two unknowns
- To define the meaning of equivalent systems of equations

You need to know

- How to represent a linear equation in two unknowns as a line in the xy-plane
- How to multiply matrices
- How to find the inverse of a 2×2 matrix

Systems of 2 × 2 equations

A system of linear equations is a set of linear equations containing the same variables.

A set of two linear equations with two variables is called a 2×2 system of linear equations,

e.g. $\begin{array}{l} a_1x + a_2y = a_3 \\ b_1x + b_2y = b_3 \end{array}$

Consistency of a system of equations

The equations $\begin{array}{l} a_1x + a_2y = a_3 \\ b_1x + b_2y = b_3 \end{array}$ can be represented by two lines in the xy-plane.

These lines may intersect, in which case there is only one set of values of x and y that satisfies the equations, i.e. there is a unique solution.

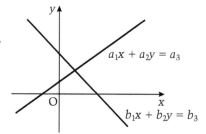

The lines may be parallel, in which case there is no solution, e.g.

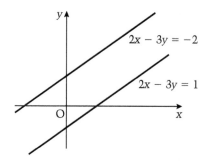

Or they may be the same line, in which case there is an infinite number of solutions, e.g.

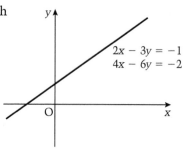

A system of equations that has either a unique solution or an infinite number of solutions is called *consistent*.

A system of equations that does not have either a unique solution or an infinite number of solutions is not consistent.

Matrix representation

The equations $\begin{array}{l} a_1 x + a_2 y = a_3 \\ b_1 x + b_2 y = b_3 \end{array}$ can be represented by a single matrix equation.

Since $\begin{pmatrix} a_1 & a_2 \\ b_1 & b_2 \end{pmatrix} \begin{pmatrix} x \\ y \end{pmatrix} = \begin{pmatrix} a_1 x + a_2 y \\ b_1 x + b_2 y \end{pmatrix}$,

we can express the equations in the form $\begin{pmatrix} a_1 & a_2 \\ b_1 & b_2 \end{pmatrix} \begin{pmatrix} x \\ y \end{pmatrix} = \begin{pmatrix} a_3 \\ b_3 \end{pmatrix}$

Using $\mathbf{A} = \begin{pmatrix} a_1 & a_2 \\ b_1 & b_2 \end{pmatrix}$, the matrix equation can be written as $\mathbf{A} \begin{pmatrix} x \\ y \end{pmatrix} = \begin{pmatrix} a_3 \\ b_3 \end{pmatrix}$

then, provided that $\mathbf{A}^{-1}$ exists, premultiplying each side by $\mathbf{A}^{-1}$ gives

$$\mathbf{A}^{-1} \mathbf{A} \begin{pmatrix} x \\ y \end{pmatrix} = \mathbf{A}^{-1} \begin{pmatrix} a_3 \\ b_3 \end{pmatrix},$$

i.e. $\begin{pmatrix} x \\ y \end{pmatrix} = \mathbf{A}^{-1} \begin{pmatrix} a_3 \\ b_3 \end{pmatrix}$

The equations have a unique solution provided that $\mathbf{A}^{-1}$ exists, i.e. provided that $|\mathbf{A}| \neq 0$

For example, the equations $\begin{array}{l} 2x - 3y = 5 \\ 4x + y = -1 \end{array}$ can be written as

$$\begin{pmatrix} 2 & -3 \\ 4 & 1 \end{pmatrix} \begin{pmatrix} x \\ y \end{pmatrix} = \begin{pmatrix} 5 \\ -1 \end{pmatrix}$$

Then if $\mathbf{A} = \begin{pmatrix} 2 & -3 \\ 4 & 1 \end{pmatrix}$, $|\mathbf{A}| = 14$ so the equations have a unique solution.

Now $\mathbf{A}^{-1} = \dfrac{1}{14} \begin{pmatrix} 1 & 3 \\ -4 & 2 \end{pmatrix}$

so $\dfrac{1}{14} \begin{pmatrix} 1 & 3 \\ -4 & 2 \end{pmatrix} \begin{pmatrix} 2 & -3 \\ 4 & 1 \end{pmatrix} \begin{pmatrix} x \\ y \end{pmatrix} = \dfrac{1}{14} \begin{pmatrix} 1 & 3 \\ -4 & 2 \end{pmatrix} \begin{pmatrix} 5 \\ -1 \end{pmatrix}$

$\Rightarrow \mathbf{I} \begin{pmatrix} x \\ y \end{pmatrix} = \dfrac{1}{14} \begin{pmatrix} 1 & 3 \\ -4 & 2 \end{pmatrix} \begin{pmatrix} 5 \\ -1 \end{pmatrix}$

$\Rightarrow \begin{pmatrix} x \\ y \end{pmatrix} = \dfrac{1}{14} \begin{pmatrix} 2 \\ -22 \end{pmatrix} = \begin{pmatrix} \frac{1}{7} \\ -\frac{11}{7} \end{pmatrix}$

Therefore $x = \frac{1}{7}$ and $y = -\frac{11}{7}$

The advantage of using matrices is that the process is mechanistic, i.e. requires no thought, so computers can easily be programmed to carry it out.

However, to solve a system of 2×2 equations by hand, it is often simpler to use the basic method of elimination or substitution.

Equivalent systems

The equations given above, i.e. $\begin{matrix} 2x - 3y = & 5 & [1] \\ 4x + y = & -1 & [2] \end{matrix}$ can be combined in several ways to give a different pair of equations.

For example [1] + [2] and [1] gives $\begin{matrix} 2x - 3y = 5 \\ 6x - 2y = 4 \end{matrix}$ and these equations have the same solution as the first set.

Any algebraic combination of equations [1] and [2] will give another set of equations with the same solution.

Two sets of equations with the same solution are called **equivalent systems**.

The aim in producing an equivalent set of equations is to make the solution easier.

For the equations [1] and [2], 3[2] + [1] gives $14x = 2$,

so the equations $\begin{matrix} 14x = 7 \\ 4x + y = -1 \end{matrix}$ have the same solution as the original pair of equations and are easier to solve.

Comparing the matrix equations of $\begin{matrix} 2x - 3y = & 5 \\ 4x + y = & -1 \end{matrix}$ and $\begin{matrix} 14x = 2 \\ 4x + y = -1 \end{matrix}$

i.e. $\begin{pmatrix} 2 & -3 \\ 4 & 1 \end{pmatrix}\begin{pmatrix} x \\ y \end{pmatrix} = \begin{pmatrix} 5 \\ -1 \end{pmatrix}$ and $\begin{pmatrix} 14 & 0 \\ 4 & 1 \end{pmatrix}\begin{pmatrix} x \\ y \end{pmatrix} = \begin{pmatrix} 2 \\ -1 \end{pmatrix}$

we can see that by replacing the first row by $(3 \times \text{row } 2 + \text{row } 1)$ of both $\begin{pmatrix} 2 & -3 \\ 4 & 1 \end{pmatrix}$ and the column vector $\begin{pmatrix} 5 \\ -1 \end{pmatrix}$ we can obtain the second matrix equation.

By placing the column vector in the matrix to get a third column, we get the **augmented matrix** $\left(\begin{matrix} 2 & -3 \\ 4 & 1 \end{matrix}\,\middle|\,\begin{matrix} 5 \\ -1 \end{matrix}\right)$

Operating on this augmented matrix ensures that whatever we do with the rows of the square matrix, we also do with the rows of the column vector, so producing an equivalent system of equations.

Calling the top row of the augmented matrix r_1 and the second row r_2, then

$3r_2 + r_1$ on the first row gives $\left(\begin{matrix} 14 & 0 \\ 4 & 1 \end{matrix}\,\middle|\,\begin{matrix} 2 \\ -1 \end{matrix}\right)$

$r_1 \div 14$ gives $\left(\begin{matrix} 1 & 0 \\ 4 & 1 \end{matrix}\,\middle|\,\begin{matrix} \frac{1}{7} \\ -1 \end{matrix}\right)$

$r_2 - 4r_1$ gives $\left(\begin{matrix} 1 & 0 \\ 0 & 1 \end{matrix}\,\middle|\,\begin{matrix} \frac{1}{7} \\ -\frac{11}{7} \end{matrix}\right)$

This gives the matrix equation $\begin{pmatrix} 1 & 0 \\ 0 & 1 \end{pmatrix}\begin{pmatrix} x \\ y \end{pmatrix} = \begin{pmatrix} \frac{1}{7} \\ -\frac{11}{7} \end{pmatrix} \Rightarrow \begin{pmatrix} x \\ y \end{pmatrix} = \begin{pmatrix} \frac{1}{7} \\ -\frac{11}{7} \end{pmatrix}$

so $y = -\frac{11}{7}$ and $x = \frac{1}{7}$

This method of solving the equations is called **_row reduction_**. The aim is to get the square matrix in the form $\begin{pmatrix} a & 0 \\ 0 & b \end{pmatrix}$

This method of solving a pair of 2×2 linear equations is clearly not as quick as using simple algebraic elimination. However, it is extended in the next topic to solve systems of 3×3 linear equations, when it does give an easier solution than purely algebraic methods. Therefore it is worth spending time practising the method on simpler 2×2 equations.

Example

Use the row reduction method to solve the equations $\begin{array}{r} 3x + y = 2 \\ 2x - 3y = 4 \end{array}$

Expressing the equations in matrix form gives $\begin{pmatrix} 3 & 1 \\ 2 & -3 \end{pmatrix}\begin{pmatrix} x \\ y \end{pmatrix} = \begin{pmatrix} 2 \\ 4 \end{pmatrix}$

Using the augmented matrix $\left(\begin{array}{cc|c} 3 & 1 & 2 \\ 2 & -3 & 4 \end{array} \right)$ gives

$3r_1 + r_2 \quad \Rightarrow \quad \left(\begin{array}{cc|c} 11 & 0 & 10 \\ 2 & -3 & 4 \end{array} \right)$

$11r_2 - 2r_1 \quad \Rightarrow \quad \left(\begin{array}{cc|c} 11 & 0 & 10 \\ 0 & -33 & 24 \end{array} \right)$

$\therefore \quad \begin{pmatrix} 11 & 0 \\ 0 & -33 \end{pmatrix}\begin{pmatrix} x \\ y \end{pmatrix} = \begin{pmatrix} 10 \\ 24 \end{pmatrix}$

$\Rightarrow \quad \begin{pmatrix} 11x \\ -33y \end{pmatrix} = \begin{pmatrix} 10 \\ 24 \end{pmatrix}$

So $\quad x = \frac{10}{11}$ and $y = -\frac{8}{11}$

Exercise 3.15

1 Determine which of the following systems of equations are consistent.

(a) $\begin{array}{r} 3x - 2y = 5 \\ 4x - y = -1 \end{array}$ (b) $\begin{array}{r} 6x - 4y = 2 \\ 3x - 2y = 1 \end{array}$ (c) $\begin{array}{r} x + y = -1 \\ x - y = -1 \end{array}$

2 Express the equations $\begin{array}{r} 5x + 3y = 8 \\ 2x - y = -4 \end{array}$ as a matrix equation.

Hence solve the equations using row reduction.

3 Solve the equations $\begin{array}{r} 6x - y = 8 \\ 2x - y = 0 \end{array}$ using row reduction on an augmented matrix.

Learning outcomes

- To investigate the consistency of a 3 × 3 system of linear equations
- To represent a set of three linear equations in three unknowns as a single matrix equation
- To reduce a matrix to row echelon form
- To solve a 3 × 3 system of linear equations using row reduction of an augmented matrix

You need to know

- That a linear equation in three unknowns can be represented as a plane in three dimensions
- How to find the determinant of a 3 × 3 matrix

Systems of 3 × 3 equations

A set of three linear equations with three variables is called a 3 × 3 system of linear equations,

e.g.
$$a_1x + a_2y + a_3z = a_4$$
$$b_1x + b_2y + b_3z = b_4$$
$$c_1x + c_2y + c_3z = c_4$$

Consistency of a system of equations

The equations $\begin{aligned} a_1x + a_2y + a_3z &= a_4 \\ b_1x + b_2y + b_3z &= b_4 \\ c_1x + c_2y + c_3z &= c_4 \end{aligned}$ can be represented by three planes.

If the planes intersect in only one point, the equations have a unique solution.

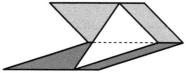

If the planes intersect in a common line, or are identical (i.e. the equations are multiples of each other), there is an infinite set of solutions.

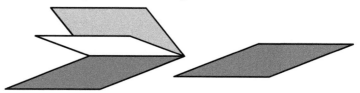

In both these cases the system of equations is consistent.

Any other configuration of the three planes will not give any solution and the equations they represent are not consistent.

Matrix representation

The equations $\begin{aligned} a_1x + a_2y + a_3z &= a_4 \\ b_1x + b_2y + b_3z &= b_4 \\ c_1x + c_2y + c_3z &= c_4 \end{aligned}$ can be expressed as the single matrix

equation $\begin{pmatrix} a_1 & a_2 & a_3 \\ b_1 & b_2 & b_3 \\ c_1 & c_2 & c_3 \end{pmatrix} \begin{pmatrix} x \\ y \\ z \end{pmatrix} = \begin{pmatrix} a_4 \\ b_4 \\ c_4 \end{pmatrix}$ as each row of the matrix multiplied by

the column vector $\begin{pmatrix} x \\ y \\ z \end{pmatrix}$ gives the left-hand side of each equation.

If $\mathbf{A} = \begin{pmatrix} a_1 & a_2 & a_3 \\ b_1 & b_2 & b_3 \\ c_1 & c_2 & c_3 \end{pmatrix}$ then, provided that $\mathbf{A}^{-1}$ exists, premultiplying each

side of the matrix equation by $\mathbf{A}^{-1}$ gives

$$\mathbf{A}^{-1}\mathbf{A}\begin{pmatrix} x \\ y \\ z \end{pmatrix} = \mathbf{A}^{-1}\begin{pmatrix} a_4 \\ b_4 \\ c_4 \end{pmatrix}, \quad \text{i.e.} \quad \mathbf{I}\begin{pmatrix} x \\ y \\ z \end{pmatrix} = \mathbf{A}^{-1}\begin{pmatrix} a_4 \\ b_4 \\ c_4 \end{pmatrix} \quad \Rightarrow \quad \begin{pmatrix} x \\ y \\ z \end{pmatrix} = \mathbf{A}^{-1}\begin{pmatrix} a_4 \\ b_4 \\ c_4 \end{pmatrix}$$

(If $\mathbf{A}^{-1}$ does not exist, the system of equations is not consistent.)

For example, the equations $\begin{array}{rcl} 4x + y &=& 3 \\ -x + 2y + z &=& 2 \\ 3x - 2y + z &=& 1 \end{array}$ can be expressed as

$$\begin{pmatrix} 4 & 1 & 0 \\ -1 & 2 & 1 \\ 3 & -2 & 1 \end{pmatrix}\begin{pmatrix} x \\ y \\ z \end{pmatrix} = \begin{pmatrix} 3 \\ 2 \\ 1 \end{pmatrix}$$

Then $\quad \mathbf{A} = \begin{pmatrix} 4 & 1 & 0 \\ -1 & 2 & 1 \\ 3 & -2 & 1 \end{pmatrix}$ and $\mathbf{A}^{-1} = \dfrac{1}{20}\begin{pmatrix} 4 & -1 & 1 \\ 4 & 4 & -4 \\ -4 & 11 & 9 \end{pmatrix}$

(This was found in Topic 3.14)

$$\therefore \quad \frac{1}{20}\begin{pmatrix} 4 & -1 & 1 \\ 4 & 4 & -4 \\ -4 & 11 & 9 \end{pmatrix}\begin{pmatrix} 4 & 1 & 0 \\ -1 & 2 & 1 \\ 3 & -2 & 1 \end{pmatrix}\begin{pmatrix} x \\ y \\ z \end{pmatrix} = \frac{1}{20}\begin{pmatrix} 4 & -1 & 1 \\ 4 & 4 & -4 \\ -4 & 11 & 9 \end{pmatrix}\begin{pmatrix} 3 \\ 2 \\ 1 \end{pmatrix}$$

$$\Rightarrow \quad \mathbf{I}\begin{pmatrix} x \\ y \\ z \end{pmatrix} = \frac{1}{20}\begin{pmatrix} 4 & -1 & 1 \\ 4 & 4 & -4 \\ -4 & 11 & 9 \end{pmatrix}\begin{pmatrix} 3 \\ 2 \\ 1 \end{pmatrix}$$

$$\Rightarrow \quad \begin{pmatrix} x \\ y \\ z \end{pmatrix} = \frac{1}{20}\begin{pmatrix} 11 \\ 16 \\ 19 \end{pmatrix} = \begin{pmatrix} \frac{11}{20} \\ \frac{4}{5} \\ \frac{19}{20} \end{pmatrix}$$

$$\therefore \quad x = \frac{11}{20}, \ y = \frac{4}{5}, \ z = \frac{19}{20}$$

Again, as with solving 2×2 linear equations, this method has the advantage of being easily programmable. However, using this method without the aid of appropriate software means finding the inverse of a 3×3 matrix, which is time-consuming and prone to mistakes.

We know that combining equations to eliminate a variable produces an equivalent system of equations. This means we can solve 3×3 linear equations using the method of row reduction of an augmented matrix, just as we did when solving 2×2 systems.

First we look at the form of an augmented matrix that we need to achieve.

Row echelon form of a matrix

The leading elements in a row of a matrix are the elements reading from left to right along the row.

A matrix is in ***row echelon form*** when each row has more leading zeroes than the row above it. (It is the number of leading zeroes in a row that matters; other elements can be zero.)

Did you know?

Echelon is the Greek word for ladder.

These matrices are in row echelon form: $\begin{pmatrix} 1 & 2 & -1 \\ 0 & 3 & 5 \\ 0 & 0 & 2 \end{pmatrix}$, $\begin{pmatrix} 0 & 4 & 0 & 0 \\ 0 & 0 & 0 & 1 \\ 0 & 0 & 0 & 0 \end{pmatrix}$

These matrices are not: $\begin{pmatrix} 1 & 2 & 1 \\ 0 & 0 & 3 \\ 0 & 0 & 2 \end{pmatrix}$, $\begin{pmatrix} 0 & 0 & 0 & 5 \\ 0 & 0 & 2 & 1 \\ 0 & 2 & 3 & 0 \end{pmatrix}$

Reduced row echelon form of a matrix

**A matrix is in *reduced row echelon form* when
each row has more leading zeroes than the row above it
and the first non-zero element in each row is 1.**

For example, $\begin{pmatrix} 1 & 4 & 0 & -2 \\ 0 & 1 & 3 & 1 \\ 0 & 0 & 0 & 1 \end{pmatrix}$ and $\begin{pmatrix} 0 & 1 & 3 & 2 \\ 0 & 0 & 0 & 1 \\ 0 & 0 & 0 & 0 \end{pmatrix}$ are in reduced row echelon form.

Using reduced row reduction to solve systems of 3 × 3 linear equations

Using the equations on page 161 again,

i.e. $\begin{pmatrix} 4 & 1 & 0 \\ -1 & 2 & 1 \\ 3 & -2 & 1 \end{pmatrix} \begin{pmatrix} x \\ y \\ z \end{pmatrix} = \begin{pmatrix} 3 \\ 2 \\ 1 \end{pmatrix}$, the augmented matrix is

$$\left(\begin{array}{ccc|c} 4 & 1 & 0 & 3 \\ -1 & 2 & 1 & 2 \\ 3 & -2 & 1 & 1 \end{array} \right)$$

We now use combinations of rows to change this to reduced row echelon form. It is important that you use combinations of rows; do not be tempted to use columns as this will not give an equivalent system of equations.

Using r_1, r_2 and r_3 to denote the rows of a matrix, we want zeroes in the leading elements in the second and third rows.

Adding $3r_2$ to r_3 gives $\left(\begin{array}{ccc|c} 4 & 1 & 0 & 3 \\ -1 & 2 & 1 & 2 \\ 3 & 4 & 4 & 7 \end{array} \right)$

Adding r_1 to $4r_2$ gives $\left(\begin{array}{ccc|c} 4 & 1 & 0 & 3 \\ 0 & 9 & 4 & 11 \\ 0 & 4 & 4 & 7 \end{array} \right)$

We now want a zero in the second element in the third row:

subtracting $4r_2$ from $9r_3$ gives $\left(\begin{array}{ccc|c} 4 & 1 & 0 & 3 \\ 0 & 9 & 4 & 11 \\ 0 & 0 & 20 & 19 \end{array} \right)$

Next, divide each row by the value of the first non-zero element in that row:

$$\begin{pmatrix} 1 & \frac{1}{4} & 0 & \bigg| & \frac{3}{4} \\ 0 & 1 & \frac{4}{9} & \bigg| & \frac{11}{9} \\ 0 & 0 & 1 & \bigg| & \frac{19}{20} \end{pmatrix}$$

This augmented matrix gives the equivalent system
$$\begin{array}{ll} x + \frac{1}{4}y = \frac{3}{4} & [1] \\ y + \frac{4}{9}z = \frac{11}{9} & [2] \\ z = \frac{19}{20} & [3] \end{array}$$

which can be easily solved using substitution, i.e. [3] in [2] gives

$$y + \left(\frac{4}{9}\right)\left(\frac{19}{20}\right) = \frac{11}{9} \;\Rightarrow\; y = \frac{220 - 76}{180} = \frac{144}{180} = \frac{4}{5}$$

then substituting the value of y into [1] gives $x + \frac{1}{5} = \frac{3}{4} \;\Rightarrow\; x = \frac{11}{20}$

Therefore the solution is $x = \frac{11}{20}, y = \frac{4}{5}, z = \frac{19}{20}$

Example

Use reduced row reduction to solve the equations
$$\begin{array}{r} 4x - y + 5z = 8 \\ 5x + 7y - 3z = 42 \\ 3x + 4y + z = 27 \end{array}$$

Starting with the augmented matrix: $\begin{pmatrix} 4 & -1 & 5 & \big| & 8 \\ 5 & 7 & -3 & \big| & 42 \\ 3 & 4 & 1 & \big| & 27 \end{pmatrix}$

$3r_2 - 5r_3 \Rightarrow \begin{pmatrix} 4 & -1 & 5 & \big| & 8 \\ 0 & 1 & -14 & \big| & -2 \\ 3 & 4 & 1 & \big| & 27 \end{pmatrix}$; $\quad 4r_3 - 3r_1 \Rightarrow \begin{pmatrix} 4 & -1 & 5 & \big| & 8 \\ 0 & 1 & -14 & \big| & -9 \\ 0 & 19 & -11 & \big| & 84 \end{pmatrix}$

$r_3 - 19r_2 \Rightarrow \begin{pmatrix} 4 & -1 & 5 & \big| & 8 \\ 0 & 1 & -14 & \big| & -9 \\ 0 & 0 & 255 & \big| & 255 \end{pmatrix}$; $r_1 \div 4, r_3 \div 255 \Rightarrow \begin{pmatrix} 1 & -\frac{1}{4} & \frac{5}{4} & \big| & 2 \\ 0 & 1 & -14 & \big| & -9 \\ 0 & 0 & 1 & \big| & 1 \end{pmatrix}$

This gives the equivalent set of equations (starting with the last row)

$$\begin{array}{ll} z = 1 & \\ y - 14z = -9 & \Rightarrow y = 5 \\ x - \frac{1}{4}y + \frac{5}{4}z = 2 & \Rightarrow x = 2 \end{array}$$

Therefore $x = 2, y = 5, z = 1$

Exercise 3.16

Use row reduction of an augmented matrix to solve the following systems of equations.

1 $\begin{array}{l} 2x - y + 3z = 8 \\ 4x + 2y - z = 13 \\ 2x + 3y - 4z = 5 \end{array}$
 2 $\begin{array}{l} x + 2y - 4z = 0 \\ 3x - y + 2z = 7 \\ 5x + y + 4z = 3 \end{array}$

3.17 Using row reduction to find an inverse matrix

Learning outcomes

- To find the multiplicative inverse of a matrix using row reduction

You need to know

- How to represent a system of linear equations in matrix form
- How to reduce a matrix to echelon form

Finding the multiplicative inverse of a matrix using row reduction

Consider the system of equations
$$3x + 2y + x = 4$$
$$x - 3y + 2z = 2$$
$$2x - 3y - z = -2$$

These can be represented by
$$\begin{pmatrix} 3 & 2 & 1 \\ 1 & -3 & 2 \\ 2 & -3 & -1 \end{pmatrix} \begin{pmatrix} x \\ y \\ z \end{pmatrix} = \begin{pmatrix} 4 \\ 2 \\ -2 \end{pmatrix}$$

$$= \begin{pmatrix} 1 & 0 & 0 \\ 0 & 1 & 0 \\ 0 & 0 & 1 \end{pmatrix} \begin{pmatrix} 4 \\ 2 \\ -2 \end{pmatrix} \quad [1]$$

We know that operating on the rows produces an equivalent system of equations.

To find the inverse of the left-hand matrix we want to reduce it to a unit matrix. So if

$$\mathbf{A} = \begin{pmatrix} 3 & 2 & 1 \\ 1 & -3 & 2 \\ 2 & -3 & -1 \end{pmatrix}, \text{ we reduce the system to } \mathbf{I}\begin{pmatrix} x \\ y \\ z \end{pmatrix} = \mathbf{A}^{-1}\begin{pmatrix} 4 \\ 2 \\ -2 \end{pmatrix}$$

Any row operation either on $\mathbf{I}$ or on $\begin{pmatrix} 4 \\ 2 \\ -2 \end{pmatrix}$ gives the same result on the right-hand side of [1],

e.g. $r_1 + r_2$ on $\mathbf{I} \Rightarrow \begin{pmatrix} 1 & 1 & 0 \\ 0 & 1 & 0 \\ 0 & 0 & 1 \end{pmatrix}\begin{pmatrix} 4 \\ 2 \\ -2 \end{pmatrix} = \begin{pmatrix} 6 \\ 2 \\ -2 \end{pmatrix}$

and $r_1 + r_2$ on $\begin{pmatrix} 4 \\ 2 \\ -2 \end{pmatrix} \Rightarrow \begin{pmatrix} 1 & 0 & 0 \\ 0 & 1 & 0 \\ 0 & 0 & 1 \end{pmatrix}\begin{pmatrix} 6 \\ 2 \\ -2 \end{pmatrix} = \begin{pmatrix} 6 \\ 2 \\ -2 \end{pmatrix}$

If we just want to calculate $\mathbf{A}^{-1}$, we can operate just on $\mathbf{A}$ and $\mathbf{I}$ using the augmented matrix:

$$\left(\begin{array}{ccc|ccc} 3 & 2 & 1 & 1 & 0 & 0 \\ 1 & -3 & 2 & 0 & 1 & 0 \\ 2 & -3 & -1 & 0 & 0 & 1 \end{array}\right)$$

We now work on the rows to reduce the left-hand side to $\mathbf{I}$.

$$r_1 + r_3 \Rightarrow \left(\begin{array}{ccc|ccc} 5 & -1 & 0 & 1 & 0 & 1 \\ 1 & -3 & 2 & 0 & 1 & 0 \\ 2 & -3 & -1 & 0 & 0 & 1 \end{array}\right);$$

$$r_2 + 2r_3 \Rightarrow \left(\begin{array}{ccc|ccc} 5 & -1 & 0 & 1 & 0 & 1 \\ 5 & -9 & 0 & 0 & 1 & 2 \\ 2 & -3 & -1 & 0 & 0 & 1 \end{array}\right);$$

$$9r_1 - r_2 \Rightarrow \left(\begin{array}{ccc|ccc} 40 & 0 & 0 & 9 & -0 & 7 \\ 5 & -9 & 0 & 0 & 1 & 2 \\ 2 & -3 & -1 & 0 & 0 & 1 \end{array}\right);$$

$$r_2 - \tfrac{1}{8}r_1 \Rightarrow \left(\begin{array}{ccc|ccc} 40 & 0 & 0 & 9 & -1 & 7 \\ 0 & -9 & 0 & -\tfrac{9}{8} & \tfrac{9}{8} & \tfrac{9}{8} \\ 2 & -3 & -1 & 0 & 0 & 1 \end{array}\right);$$

$$r_3 - \tfrac{1}{20}r_1 \Rightarrow \left(\begin{array}{ccc|ccc} 40 & 0 & 0 & 9 & -1 & 7 \\ 0 & -9 & 0 & -\tfrac{9}{8} & \tfrac{9}{8} & \tfrac{9}{8} \\ 0 & -3 & -1 & -\tfrac{9}{20} & \tfrac{1}{20} & \tfrac{13}{20} \end{array}\right);$$

$$3r_3 - r_2 \Rightarrow \left(\begin{array}{ccc|ccc} 40 & 0 & 0 & 9 & -1 & 7 \\ 0 & -9 & 0 & -\tfrac{9}{8} & \tfrac{9}{8} & \tfrac{9}{8} \\ 0 & 0 & -3 & -\tfrac{9}{40} & -\tfrac{39}{40} & \tfrac{33}{40} \end{array}\right);$$

$$\begin{array}{ccc} \tfrac{r_1}{40}, & \tfrac{r_2}{-9}, & \tfrac{r_3}{-3} \end{array} \Rightarrow \left(\begin{array}{ccc|ccc} 1 & 0 & 0 & \tfrac{9}{40} & -\tfrac{1}{40} & \tfrac{7}{40} \\ 0 & 1 & 0 & \tfrac{1}{8} & -\tfrac{1}{8} & -\tfrac{1}{8} \\ 0 & 0 & 1 & \tfrac{3}{40} & \tfrac{13}{40} & -\tfrac{11}{40} \end{array}\right)$$

We have now reduced the system in [1] to

$$\begin{pmatrix} 1 & 0 & 0 \\ 0 & 1 & 0 \\ 0 & 0 & 1 \end{pmatrix}\begin{pmatrix} x \\ y \\ z \end{pmatrix} = \begin{pmatrix} \tfrac{9}{40} & -\tfrac{1}{40} & \tfrac{7}{40} \\ \tfrac{1}{8} & -\tfrac{1}{8} & -\tfrac{1}{8} \\ \tfrac{3}{40} & \tfrac{13}{40} & -\tfrac{11}{40} \end{pmatrix}\begin{pmatrix} 4 \\ 2 \\ -2 \end{pmatrix}$$

$$\therefore \; \mathbf{A}^{-1} = \begin{pmatrix} \tfrac{9}{40} & -\tfrac{1}{40} & \tfrac{7}{40} \\ \tfrac{1}{8} & -\tfrac{1}{8} & -\tfrac{1}{8} \\ \tfrac{3}{40} & \tfrac{13}{40} & -\tfrac{11}{40} \end{pmatrix} = \frac{1}{40}\begin{pmatrix} 9 & -1 & 7 \\ 5 & -5 & -5 \\ 3 & 13 & -11 \end{pmatrix}$$

This method of finding an inverse of a matrix by row reduction has advantages over the method using cofactors because it simplifies the arithmetic, and so mistakes are less likely. However, it is sensible to check that your calculated inverse multiplied by the original matrix does give **I**.

Using row reduction is also a quicker method for showing that a matrix is singular as it will produce a row of zeroes, proving that $|\mathbf{A}| = 0$

Exercise 3.17

1 Use row reduction to show that the matrix $\begin{pmatrix} 1 & 3 & 1 \\ 3 & 7 & 2 \\ 2 & 4 & 1 \end{pmatrix}$ is singular.

2 Find the inverse of each of the following matrices.

(a) $\begin{pmatrix} 4 & 1 & 1 \\ 2 & 0 & 2 \\ 1 & -2 & 1 \end{pmatrix}$ (b) $\begin{pmatrix} 2 & 1 & 0 \\ 5 & 5 & 1 \\ 1 & -2 & 1 \end{pmatrix}$ (c) $\begin{pmatrix} 2 & -1 & 4 \\ 2 & -1 & 5 \\ 1 & 2 & 4 \end{pmatrix}$

3.18 Differential equations

Learning outcomes

- To explain differential equations as mathematical models
- To formulate differential equations of the form $\dfrac{dy}{dx} - ky = f(x)$ where k is a function of x or a constant

You need to know

- The meaning of a first and second derivative
- The basic facts about integration
- The derivatives of standard functions
- How to differentiate a product of functions
- How to differentiate implicit functions

Did you know?

Newton's name keeps appearing in the study of mathematics. Sir Isaac Newton (1643−1727) was a prolific mathematician. He was also arguably the greatest scientist the world has known.

Differential equations

A **differential equation** connects an unknown function and its derivatives.

The **order of a differential equation** is the highest derivative contained in the equation.

For example, $\dfrac{dy}{dx} - xy = x^2$ is a first order differential equation and

$\dfrac{d^2y}{dx^2} + y\dfrac{dy}{dx} - xy = 0$ is a second order differential equation.

Models

Unlike many topics in mathematics which find real-world applications some time after their development, the formulation of differential equations comes directly from the need to describe real-world phenomena mathematically.

A differential equation is a mathematical description of a real-world phenomenon. It is used to predict results and it is called a **mathematical model**.

How good the model is depends on how closely the results it predicts are to measured results from the real-world phenomenon.

There are many well-known equations that are extremely good models. For example, Newton's laws of motion are a set of equations that describe the relationship between the forces acting on a body and the motion of the body. These are accurate enough to be used to determine the forces needed to place satellites in orbit.

Solution of differential equations

The solution of a differential equation gives an equation connecting the variables without any derivatives involved. If the differential equation is a first order equation, solving it involves one integration operation so it will include one unknown constant. When the differential equation is a second order equation, two integration operations are needed to solve it so the solution will involve two unknown constants.

There is an enormous number of different types of differential equation and many of them cannot be solved.

In Unit 1 we covered the solution of a first order differential equation with separable variables. In the remaining topics in Unit 2 we look at the solution of two more types of differential equation.

First, we look at how a particular type of differential equation can arise.

Formulation of differential equations of the form
$\frac{dy}{dx} - ky = f(x)$ where $k = h(x)$

We know that, if u and v are functions of x, then $\frac{d}{dx}(uv) = u\frac{dv}{dx} + v\frac{du}{dx}$

We also know that $\frac{d}{dx}(yg(x)) = g(x)\frac{dy}{dx} + yg'(x)$

for example, $\frac{d}{dx}(y\sin x) = \sin x\frac{dy}{dx} + y\cos x$

Therefore, given the differential equation $\sin x\frac{dy}{dx} + y\cos x = 2x$, we can recognise the left-hand side as the differential of $y\sin x$, and so solve the equation by integrating both sides.

Hence $\quad \sin x\frac{dy}{dx} + y\cos x = 2x$

$\Rightarrow \qquad\qquad y\sin x = x^2 + A \quad$ where A is an unknown constant.

This type of differential equation is called an **exact differential equation**.

If the equation $\sin x\frac{dy}{dx} + y\cos x = 2x$ is divided by $\sin x$ $(\sin x \neq 0)$ it becomes

$$\frac{dy}{dx} + y\cot x = 2x\,\text{cosec}\,x$$

so it is of the form $\quad \frac{dy}{dx} - ky = f(x) \quad$ where $k = h(x)$ but the left-hand side is not now the derivative of a product. Before finding ways of solving differential equations of this form, we solve a few exact differential equations that have not been simplified.

The solution of a differential equation containing unknown constants of integration is called the **general solution**. To evaluate constants of integration we need to know initial values of x and y. These are called **boundary conditions**.

Example

Find the general solution of the differential equation $x\frac{dy}{dx} + y = \frac{1}{x}$

The left-hand side is the derivative of xy,

therefore the general solution is $xy = \ln|x| + A$

Exercise 3.18

Find the general solution of each of the following differential equations.

1 $e^x\frac{dy}{dx} + ye^x = 2x$
2 $x^2\frac{dy}{dx} + 2xy = \cos x$

3 $t\frac{dv}{dt} + v = t^2$

Learning outcomes

- To solve differential equations of the form $\frac{dy}{dx} - yg(x) = f(x)$ using an integrating factor

You need to know

- The formula for differentiating a product
- The chain rule
- How to solve a differential equation with separable variables
- The integrals of standard functions

Integrating factors

We know that if the left-hand side of a differential equation has the form $g(x)\frac{dy}{dx} + yg'(x)$ we recognise that this is the derivative of the product $yg(x)$.

However, when this is not the case, it is possible to multiply by a function of x that will give the derivative of the product $yg(x)$.

This function is called an **integrating factor**, and we will denote it by I.

Consider an equation of the form $\frac{dy}{dx} + Gy = F$ where both G and F are functions of x.

Now multiply the equation by I where I is a function of x,

i.e. $\quad I\frac{dy}{dx} + (y)(GI) = FI$ $\qquad$ [1]

We want to find I such that $\frac{d}{dx}(Iy) = I\frac{dy}{dx} + (y)(GI)$

Comparing the left-hand side of [1] with $u\frac{dv}{dx} + v\frac{du}{dx}$ gives

$$u = I, \ \frac{dv}{dx} = \frac{dy}{dx} \quad \text{and} \quad v = y, \ \frac{du}{dx} = GI$$

Using the chain rule, i.e. $\frac{dI}{dx} = \frac{dI}{du} \times \frac{du}{dx}$, gives $\frac{dI}{dx} = 1 \times GI = GI$

Now $\frac{dI}{dx} = GI$ is a first order differential equation whose variables are separable.

Therefore $\quad \int \frac{1}{I}\,dI = \int G\,dx$

$\Rightarrow \qquad \ln I = \int G\,dx$

$\qquad\qquad\qquad \Rightarrow \ \boldsymbol{I = e^{\int G\,dx}}$

So $I = e^{\int G\,dx}$ is an integrating factor for the expression $\frac{dy}{dx} + Gy$

Assuming that $\int G\,dx$ can be found (it cannot always, but can for any equations you meet in the examination), then

$\frac{dy}{dx} + Gy = F \quad \Rightarrow \quad I\frac{dy}{dx} + (y)(GI) = FI$

$\qquad\qquad \Rightarrow \ \int\left(I\frac{dy}{dx} + (y)(GI)\right)dx = \int IF\,dx$

$\qquad\qquad \Rightarrow \ Iy = \int IF\,dx$

Both I and F are functions of x, so the integral on the right-hand side can be found for any equations you meet in the examination.

Example

Find the general solution of the differential equation $\dfrac{dy}{dx} - \dfrac{2y}{x} = x^2 e^x$

$\dfrac{dy}{dx} - \dfrac{2y}{x} = x^2 e^x \;\Rightarrow\; \dfrac{dy}{dx} + y\left(-\dfrac{2}{x}\right) = x^2 e^x$

$\therefore$ the integrating factor I is $e^{\int -\frac{2}{x}dx} = e^{-2\ln x}$

$\qquad\qquad\qquad\qquad\qquad = e^{\ln(x^{-2})} = x^{-2}$

Multiplying both sides of $\dfrac{dy}{dx} - \dfrac{2y}{x} = x^2 e^x$ by x^{-2} gives

$\qquad \dfrac{1}{x^2}\dfrac{dy}{dx} - \dfrac{2y}{x^3} = e^x$

$\Rightarrow \quad \dfrac{y}{x^2} = \displaystyle\int e^x\, dx$

$\qquad\qquad = e^x + A$

$\Rightarrow \quad y = x^2 e^x + Ax^2$

Example

Find the solution of the differential equation $t\dfrac{dv}{dt} - v + t^2 e^{-t} = 0$
given that $v = 0$ when $t = 1$

First, rearrange the equation so that it is in the form $\dfrac{dv}{dt} + vg(t) = f(t)$

$\qquad t\dfrac{dv}{dt} - v + t^2 e^{-t} = 0 \;\Rightarrow\; \dfrac{dv}{dt} + v\left(-\dfrac{1}{t}\right) = -te^{-t}$

$\therefore \;\; I = e^{\int -\frac{1}{t}dt} = e^{-\ln t} = \dfrac{1}{t}$

$\therefore \;\; \dfrac{1}{t}\dfrac{dv}{dt} - v\left(\dfrac{1}{t^2}\right) = -e^{-t}$

$\Rightarrow \;\; v\left(\dfrac{1}{t}\right) = \displaystyle\int -e^{-t}\, dt$

$\Rightarrow \;\; v = te^{-t} + At$

$v = 0$ when $t = 1$ gives $0 = e^{-1} + A$ so $A = -e^{-1}$

$\therefore \;\; v = te^{-t} - te^{-1}$

Exercise 3.19

1 Find the general solution of

(a) $\dfrac{dy}{dx} + 3y = x$

(b) $\sin x\dfrac{dy}{dx} + y\cos x = 1$

2 Find the solution of $\dfrac{dy}{dx} + xy = x$ given $y = 0$ when $x = 0$

3 Solve the equation $\theta\dfrac{dy}{d\theta} - y = \theta^2\cos\theta$ given $y = 0$ when $\theta = \dfrac{\pi}{2}$

Summary of solutions of first order differential equations

There are many differential equations involving $\dfrac{dy}{dx}$ that cannot be solved to give a direct relationship between x and y. The table shows methods that can be used for some first order differential equations that can be solved. All of these methods rely on recognising the standard integrals, so you need to know all of these from Unit 1 as well as from Unit 2.

Form of equation	Method of solution
$\dfrac{dy}{dx} = f(x)$	Recognise the function of which the differential is $f(x)$ or use a substitution to simplify $f(x)$ or use partial fractions (for a rational function) or use integration by parts (for a product of functions).
$\dfrac{dy}{dx} = g(y)f(x)$	Separate the variables to give $\dfrac{1}{g(y)}\dfrac{dy}{dx} = f(x)$ then $\displaystyle\int \dfrac{1}{g(y)}\, dy = \int f(x)\, dx$
$g(x)\dfrac{dy}{dx} + yg'(x) = f(x)$	Recognise the left-hand side as the differential of $yg(x)$ giving $$yg(x) = \int f(x)\, dx$$ (You may need to rearrange the equation to give this form.) This form is called an exact differential equation.
$\dfrac{dy}{dx} + yg(x) = f(x)$	When rearrangement does not give an exact differential equation, multiply throughout by the integrating factor $$I = e^{\int g(x)\, dx}$$ This then gives an exact differential equation.

Example

Solve the following differential equations.

(a) $(1 + x^2)\dfrac{dy}{dx} = 2x$

(b) $(1 + x^2)\dfrac{dy}{dx} = 2y$

(c) $(1 + x^2)\dfrac{dy}{dx} + 2xy = 0$

(d) $x\dfrac{dy}{dx} + 2x^2y = 4xe^{-x^2}$

(a) $(1 + x^2)\dfrac{dy}{dx} = 2x$

This equation contains no term involving y, so first rearrange it to isolate $\dfrac{dy}{dx}$.

$\dfrac{dy}{dx} = \dfrac{2x}{1 + x^2} \quad \Rightarrow \quad y = \ln|1 + x^2| + c$

(b) $(1 + x^2)\dfrac{dy}{dx} = 2y$

The variables in this equation can be separated to give

$\dfrac{1}{y}\dfrac{dy}{dx} = \dfrac{2}{1 + x^2}$

$\Rightarrow \quad \displaystyle\int \dfrac{1}{y}\, dy = \int \dfrac{2}{1 + x^2}\, dx \quad \Rightarrow \quad \ln|y| = 2\tan^{-1}x + c$

(c) $(1 + x^2)\dfrac{dy}{dx} + 2xy = 0$

$2x$ is the differential of $(1 + x)^2$ so this is an exact differential equation giving

$y(1 + x^2) = c$

(The variables in this differential equation can be separated but recognition of an exact differential equation gives a quicker and neater solution.)

(d) $x\dfrac{dy}{dx} + 2x^2y = 4xe^{-x^2}$

The left-hand side is not exact and the variables cannot be separated but we can rearrange the equation to the form $\dfrac{dy}{dx} + yg(x) = f(x)$ and use an integrating factor:

$\dfrac{dy}{dx} + 2xy = 4e^{-x^2}$

then $\quad I = e^{\int 2x\, dx} \quad \Rightarrow \quad I = e^{x^2}$

$\therefore \quad e^{x^2}\dfrac{dy}{dx} + 2xe^{x^2}y = 4 \quad \Rightarrow \quad e^{x^2}y = 4x + c$

Exercise 3.20

Solve each of the following differential equations.

1 $\sin x \dfrac{dy}{dx} + y\cos x = \tan x$

2 $x\dfrac{dy}{dx} = \dfrac{2x}{1 - x^2}$

3 $x\dfrac{dy}{dx} - y = 3x^3$

4 $\dfrac{dy}{dx} + y = 0$

3.21 Differential equations of the form $a\dfrac{d^2y}{dx^2} + b\dfrac{dy}{dx} + cy = 0$

Learning outcomes

- To solve equations of the form
  $a\dfrac{d^2y}{dx^2} + b\dfrac{dy}{dx} + cy = 0$
  where $a, b, c \in \mathbb{R}$

You need to know

- The relationship between the roots and the coefficients of a quadratic equation
- The meaning of conjugate complex numbers

Formulation of a second order differential equation

When an equation $y = f(x)$ contains two unknown constants, A and B, differentiating twice gives $\dfrac{dy}{dx} = f'(x)$ and $\dfrac{d^2y}{dx^2} = f''(x)$

These two equations, together with the original equation, can be used to eliminate A and B, giving a second order differential equation.

We now look at three types of the equation $y = f(x)$ containing unknown constants A and B, all of which give rise to an equation of the form

$a\dfrac{d^2y}{dx^2} + b\dfrac{dy}{dx} + cy = 0$ where a, b and c are real constants.

A differential equation of the form $a\dfrac{d^2y}{dx^2} + b\dfrac{dy}{dx} + cy = 0$ is called a *linear second order differential equation.*

The equation $y = Ae^{\alpha x} + Be^{\beta x}$

Consider the equation $y = Ae^{3x} + Be^{4x}$

$$\frac{dy}{dx} = 3Ae^{3x} + 4Be^{4x} = 3y + Be^{4x}$$

$$\frac{d^2y}{dx^2} = 3\frac{dy}{dx} + 4Be^{4x} = 3\frac{dy}{dx} + 4\left(\frac{dy}{dx} - 3y\right)$$

$$\Rightarrow \quad \frac{d^2y}{dx^2} - (3+4)\frac{dy}{dx} + (3)(4)\,y = 0$$

and the coefficients of this equation are the roots of the quadratic equation $u^2 - 7u + 12 = 0$.

Therefore working backwards, by using the coefficients of

$\dfrac{d^2y}{dx^2} - 7\dfrac{dy}{dx} + 12y = 0$ to give the quadratic equation

$u^2 - 7u + 12 = 0 \quad \Rightarrow \quad (u-3)(u-4) = 0$ then the roots of this equation give the general solution as $y = Ae^{3x} + Be^{4x}$

Now consider the general case, i.e. $y = Ae^{\alpha x} + Be^{\beta x}$

$$\frac{dy}{dx} = \alpha Ae^{\alpha x} + \beta Be^{\beta x}$$

and $$\frac{d^2y}{dx^2} = \alpha^2 Ae^{\alpha x} + \beta^2 Be^{\beta x}$$

Eliminating A and B from these two equations gives

$$\frac{d^2y}{dx^2} - (\alpha + \beta)\frac{dy}{dx} + \alpha\beta y = 0$$

Then the coefficients of this equation give the quadratic equation

$$u^2 - (\alpha + \beta)u + \alpha\beta = 0$$

whose roots are α and β.

In general, the quadratic equation $au^2 + bu + c = 0$, formed from the differential equation $a\dfrac{d^2y}{dx^2} + b\dfrac{dy}{dx} + cy = 0$ is called the *auxiliary equation*.

When the auxiliary equation $au^2 + bu + c = 0$ has real distinct roots α and β, the general solution of the differential equation

$$a\dfrac{d^2y}{dx^2} + b\dfrac{dy}{dx} + cy = 0 \text{ can be quoted as}$$

$$y = Ae^{\alpha x} + Be^{\beta x}$$

Example

Find the general solution of the differential equation

$$2\dfrac{d^2y}{dx^2} - 3\dfrac{dy}{dx} + y = 0$$

The auxiliary equation is $2u^2 - 3u + 1 = 0$

$$2u^2 - 3u + 1 = 0 \quad \Rightarrow \quad (2u - 1)(u - 1) = 0 \quad \Rightarrow \quad u = \tfrac{1}{2} \text{ or } u = 1$$

Therefore the general solution of $2\dfrac{d^2y}{dx^2} - 3\dfrac{dy}{dx} + y = 0$

is $y = Ae^{\frac{1}{2}x} + Be^{x}$

The equation $y = (A + Bx)e^{\alpha x}$

Consider the equation $y = (A + Bx)e^{3x}$

$$\dfrac{dy}{dx} = 3(A + Bx)e^{3x} + Be^{3x} = 3y + Be^{3x}$$

$$\dfrac{d^2y}{dx^2} = 3\dfrac{dy}{dx} + 3Be^{3x} = 3\dfrac{dy}{dx} + 3\left(\dfrac{dy}{dx} - 3y\right)$$

$$\Rightarrow \quad \dfrac{d^2y}{dx^2} - (3 + 3)\dfrac{dy}{dx} + (3 \times 3)\,y = 0$$

This time the auxiliary quadratic equation is $u^2 - (3 + 3)u + (3 \times 3) = 0$ and it has a repeated root of 3.

Now consider the general case, i.e. $y = (A + Bx)e^{\alpha x}$

$$\dfrac{dy}{dx} = \alpha y + Be^{\alpha x}$$

$$\dfrac{d^2y}{dx^2} = \alpha\dfrac{dy}{dx} + \alpha Be^{\alpha x} = \alpha\dfrac{dy}{dx} + \alpha\left(\dfrac{dy}{dx} - \alpha y\right)$$

$$\Rightarrow \quad \dfrac{d^2y}{dx^2} - 2\alpha\dfrac{dy}{dx} + \alpha^2 y = 0$$

i.e. $y = (A + Bx)e^{\alpha x} \quad \Rightarrow \quad \dfrac{d^2y}{dx^2} - 2\alpha\dfrac{dy}{dx} + \alpha^2 y = 0$

> When the auxiliary equation $au^2 + bu + c = 0$ has a repeated root α, the general solution of the differential equation
>
> $$a\frac{d^2y}{dx^2} + b\frac{dy}{dx} + cy = 0 \text{ can be quoted as}$$
>
> $$y = (A + Bx)e^{\alpha x}$$

Example

Find the general solution of the differential equation

$$4\frac{d^2y}{dx^2} + 12\frac{dy}{dx} + 9y = 0$$

The auxiliary equation is $4u^2 + 12u + 9 = 0$

$$\Rightarrow \qquad (2u + 3)^2 = 0$$

This equation has a repeated root equal to $-\frac{3}{2}$

so the general solution of $4\dfrac{d^2y}{dx^2} + 12\dfrac{dy}{dx} + 9y = 0$

is $y = (A + Bx)e^{-\frac{3}{2}x}$

The equation $y = e^{\alpha x}(A\cos\beta x + B\sin\beta x)$

Consider the equation $y = e^{2x}(A\cos 3x + B\sin 3x)$

$$\frac{dy}{dx} = 2e^{2x}(A\cos 3x + B\sin 3x) + e^{2x}(-3A\sin 3x + 3B\cos 3x)$$

$$= 2y + e^{2x}(-3A\sin 3x + 3B\cos 3x)$$

$$\frac{d^2y}{dx^2} = 2\frac{dy}{dx} + 2e^{2x}(-3A\sin 3x + 3B\cos 3x) + e^{2x}(-9A\cos 3x - 9B\sin 3x)$$

$$= 2\frac{dy}{dx} + 2\left(\frac{dy}{dx} - 2y\right) - 9y$$

i.e. $\dfrac{d^2y}{dx^2} - 4\dfrac{dy}{dx} + 13y = 0$

The auxiliary equation is $u^2 - 4u + 13 = 0$ and the roots of this equation are the conjugate complex numbers $\dfrac{4 \pm \sqrt{-36}}{2} = 2 \pm 3i$

Now consider the general case $y = e^{\alpha x}(A\cos\beta x + B\sin\beta x)$

$$\frac{dy}{dx} = \alpha e^{\alpha x}(A\cos\beta x + B\sin\beta x) + e^{\alpha x}(-\beta A\sin\beta x + \beta B\cos\beta x)$$

$$= \alpha y + e^{\alpha x}(-\beta A\sin\beta x + \beta B\cos\beta x)$$

$$\frac{d^2y}{dx^2} = \alpha\frac{dy}{dx} + \alpha\left(\frac{dy}{dx} - \alpha y\right) - \beta^2 y$$

$$\Rightarrow \frac{d^2y}{dx^2} - 2\alpha\frac{dy}{dx} + (\alpha^2 + \beta^2)y = 0$$

The auxiliary equation is $u^2 - 2\alpha u + (\alpha^2 + \beta^2)y = 0$ and the roots of this equation are $\dfrac{2\alpha \pm \sqrt{4\alpha^2 - 4(\alpha^2 + \beta^2)}}{2} = \alpha \pm i\beta$

When the auxiliary equation $au^2 + bu + c = 0$ has complex roots $\alpha \pm i\beta$, the general solution of the differential equation

$$a\frac{d^2y}{dx^2} + b\frac{dy}{dx} + cy = 0 \text{ can be quoted as}$$

$$y = e^{\alpha x}(A\cos\beta x + B\sin\beta x)$$

Note that if the roots of the auxiliary equation are purely imaginary, i.e. $\alpha = 0$,

then $y = A\cos\beta x + B\sin\beta x$

Example

Find the general solution of the differential equation

$$\frac{d^2y}{dx^2} + 4\frac{dy}{dx} + 5y = 0$$

The auxiliary equation is

$$u^2 + 4u + 5 = 0 \quad \Rightarrow \quad u = \frac{-4 \pm \sqrt{16 - 20}}{2}$$

$$= -2 \pm i$$

Therefore the general solution is $y = e^{-2x}(A\cos x + B\sin x)$

Summary

The general solution of the differential equation

$$a\frac{d^2y}{dx^2} + b\frac{dy}{dx} + cy = 0$$

depends on the nature of the roots of the auxiliary equation

$$au^2 + bu + c = 0$$

If the roots are α and β then

- when α and β are real and distinct, $y = Ae^{\alpha x} + Be^{\beta x}$
- when $\alpha = \beta$, $y = (A + Bx)e^{\alpha x}$
- when α and β are complex conjugate numbers, $y = e^{\alpha x}(A\cos\beta x + B\sin\beta x)$

Exercise 3.21

Find the general solution of each differential equation.

1 $\dfrac{d^2y}{dx^2} - 7\dfrac{dy}{dx} + 12y = 0$ **2** $\dfrac{d^2y}{dx^2} - 2\dfrac{dy}{dx} + 5y = 0$

3 $\dfrac{d^2y}{dx^2} - 4y = 0$ **4** $\dfrac{d^2y}{dx^2} + 4y = 0$

5 $\dfrac{d^2y}{dx^2} - 7\dfrac{dy}{dx} = 0$ **6** $\dfrac{d^2y}{dx^2} - 6\dfrac{dy}{dx} + 9y = 0$

3.22 The particular integral 1

Learning outcomes

- To solve differential equations of the form
$$a\frac{d^2y}{dx^2} + b\frac{dy}{dx} + cy = f(x)$$
where f(x) is a polynomial

You need to know

- How to find the general solution of the differential equation
$$a\frac{d^2y}{dx^2} + b\frac{dy}{dx} + cy = 0$$

The particular integral when f(x) is a polynomial or a constant

Consider the differential equation $\dfrac{d^2y}{dx^2} + 4\dfrac{dy}{dx} + 3y = 6x^2 + x + 2$

The polynomial on the right-hand side suggests that a polynomial of the form $y = ax^2 + bx + c$ might be a solution of the differential equation. We call this a **trial solution**.

We can test this by differentiation, to see if values of a, b and c exist so that $y = ax^2 + bx + c$ is a solution,

i.e. $y = ax^2 + bx + c \Rightarrow \dfrac{dy}{dx} = 2ax + b$ and $\dfrac{d^2y}{dx^2} = 2a$

Substituting into the left-hand side of the differential equation gives

$$2a + 4(2ax + b) + 3(ax^2 + bx + c) = 6x^2 + x + 2$$
$$\Rightarrow \quad 3ax^2 + (8a + 3b)x + (2a + 4b + 3c) = 6x^2 + x + 2$$

Comparing coefficients gives $a = 2$, $b = -5$ and $c = 6$

Therefore $y = 2x^2 - 5x + 6$ is a solution of the equation
$$\frac{d^2y}{dx^2} + 4\frac{dy}{dx} + 3y = 6x^2 + x + 2$$

However, $y = 2x^2 - 5x + 6$ cannot be the full solution because it does not contain any constants of integration.

The function $2x^2 - 5x + 6$ is called the **particular integral**.

The general solution of $a\dfrac{d^2y}{dx^2} + b\dfrac{dy}{dx} + cy = f(x)$ where f(x) is a polynomial

We have found a solution of the equation
$$\frac{d^2y}{dx^2} + 4\frac{dy}{dx} + 3y = 6x^2 + x + 2$$

and we can find the general solution by first solving the simpler differential equation
$$\frac{d^2y}{dx^2} + 4\frac{dy}{dx} + 3y = 0$$

The auxiliary equation is
$$u^2 + 4u + 3 = 0 \quad \Rightarrow \quad (u + 3)(u + 1) = 0 \quad \Rightarrow \quad u = -3 \text{ or } -1$$
$$\therefore \quad y = Ae^{-3x} + Be^{-x}$$

Adding $Ae^{-3x} + Be^{-x}$ to the particular integral gives
$$y = Ae^{-3x} + Be^{-x} + 2x^2 - 5x + 6$$

and we can show that this is the general solution of
$$\frac{d^2y}{dx^2} + 4\frac{dy}{dx} + 3y = 6x^2 + x + 2:$$

$$\left. \begin{aligned} y &= Ae^{-3x} + Be^{-x} + 2x^2 - 5x + 6 \\ \frac{dy}{dx} &= -3Ae^{-3x} - Be^{-x} + 4x - 5 \\ \frac{d^2y}{dx^2} &= 9Ae^{-3x} + Be^{-x} + 4 \end{aligned} \right\} \quad \frac{d^2y}{dx^2} + 4\frac{dy}{dx} + 3y = 6x^2 + x + 2$$

$Ae^{-3x} + Be^{-x}$ is called the **complementary function**, and we have found the general solution of the given differential equation by adding the complementary function and the particular integral.

For any differential equation of the form $a\dfrac{d^2y}{dx^2} + b\dfrac{dy}{dx} + cy = f(x)$

where f(x) is a polynomial, the general solution is given by
y = (complementary function) + (particular integral)

where the complementary function is the solution of $a\dfrac{d^2y}{dx^2} + b\dfrac{dy}{dx} + cy = 0$

and the particular integral is a general polynomial of the same order as f(x) and whose coefficients can be found by differentiation and substitution into the given differential equation.

Example

Find the general solution of the equation $\dfrac{d^2y}{dx^2} + 2\dfrac{dy}{dx} + 5y = 7x - 1$

First find the particular integral: try $y = ax + b$

$\Rightarrow \dfrac{dy}{dx} = a$ and $\dfrac{d^2y}{dx^2} = 0$

Substituting into the given differential equation gives
$2a + 5(ax + b) = 7x - 1$

Comparing coefficients gives $a = \frac{7}{5}$ and $b = -\frac{19}{25}$

Therefore $y = \frac{7}{5}x - \frac{19}{25}$ is the particular integral.

Next find the complementary function:
the auxiliary equation is $u^2 + 2u + 5 = 0$

$\Rightarrow u = \dfrac{-2 \pm \sqrt{4 - 20}}{2} = -1 \pm 2i$

So the complementary function is
$y = e^{-x}(A\cos 2x + B\sin 2x)$

Therefore the general solution is
$y = e^{-x}(A\cos 2x + B\sin 2x) + \frac{7}{5}x - \frac{19}{25}$

Exercise 3.22

Find the general solution of each differential equation.

1 $\dfrac{d^2y}{dx^2} + \dfrac{dy}{dx} + y = 1 + x$ **2** $\dfrac{d^2y}{dx^2} - 9y = x^2 - 2$

Learning outcomes

- To solve differential equations of the form
$$a\frac{d^2y}{dx^2} + b\frac{dy}{dx} + cy = f(x)$$
where f(x) is a trigonometric function

You need to know

- How to find the general solution of the differential equation
$$a\frac{d^2y}{dx^2} + b\frac{dy}{dx} + cy = 0$$

The particular integral when f(x) is a trigonometric function

Consider the differential equation $\dfrac{d^2y}{dx^2} + 4\dfrac{dy}{dx} + 3y = 2\cos x - 3\sin x$

The function on the right-hand side suggests that a function of the form $y = p\cos x + q\sin x$ might be a solution of the differential equation. Using this as a trial solution, we can differentiate it to find out if values of p and q exist so that $y = p\cos x + q\sin x$ is a solution.

$$y = p\cos x + q\sin x$$

$$\frac{dy}{dx} = -p\sin x + q\cos x$$

$$\frac{d^2y}{dx^2} = -p\cos x - q\sin x$$

Substituting these expressions into the given differential equation gives

$$(-p\cos x - q\sin x) + 4(-p\sin x + q\cos x) + 3(p\cos x + q\sin x)$$
$$= 2\cos x - 3\sin x$$

$$\Rightarrow \quad (2p + 4q)\cos x + (-4p + 2q)\sin x = 2\cos x - 3\sin x$$

$$\Rightarrow \quad \left.\begin{array}{c} p + 2q = 1 \\ -4p + 2q = -3 \end{array}\right\} \quad \Rightarrow \quad p = \tfrac{4}{5} \text{ and } q = \tfrac{1}{10}$$

$$\therefore \quad y = \tfrac{4}{5}\cos x + \tfrac{1}{10}\sin x \text{ is a solution of the given differential equation}$$

and so $\tfrac{4}{5}\cos x + \tfrac{1}{10}\sin x$ is the particular integral.

When f(x) is any combination of cos vx and/or sin vx we use $y = p\cos vx + q\sin vx$ as the trial solution. For example, if f(x) = $3\sin 4x$, we use $y = p\cos 4x + q\sin 4x$

The general solution when
$$a\frac{d^2y}{dx^2} + b\frac{dy}{dx} + cy = \text{a trigonometric function of } x$$

For the differential equation $\dfrac{d^2y}{dx^2} + 4\dfrac{dy}{dx} + 3y = 2\cos x - 3\sin x$ the

complementary function is the general solution of $\dfrac{d^2y}{dx^2} + 4\dfrac{dy}{dx} + 3y = 0$

The auxiliary equation is

$$u^2 + 4u + 3 = 0 \quad \Rightarrow \quad (u + 3)(u + 1) = 0 \quad \Rightarrow \quad u = -3 \text{ or } -1$$

Therefore the complementary function is $Ae^{-3x} + Be^{-x}$

The general solution of the given differential equation is therefore

$$y = Ae^{-3x} + Be^{-x} + \tfrac{4}{5}\cos x + \tfrac{1}{10}\sin x$$

For any differential equation of the form $a\dfrac{d^2y}{dx^2} + b\dfrac{dy}{dx} + cy = f(x)$

where f(x) is a combination of sines and cosines of the same angle,
the general solution is given by
y = (complementary function) + (particular integral)
where the complementary function is the solution of

$a\dfrac{d^2y}{dx^2} + b\dfrac{dy}{dx} + cy = 0$ and the particular integral is $p \cos ux + q \sin ux$

and where p and q are constants which can be found by differentiation
and substitution into the given differential equation.

Example

Find the general solution of the differential equation

$$\dfrac{d^2y}{dx^2} - 6\dfrac{dy}{dx} + 9y = 4\sin 2x$$

Using $y = p\cos 2x + q\sin 2x$ as the trial solution gives

$$\dfrac{dy}{dx} = -2p\sin 2x + 2q\cos 2x$$

$$\dfrac{d^2y}{dx^2} = -4p\cos 2x - 4q\sin 2x$$

Substituting into the given differential equation gives

$(-4p\cos 2x - 4q\sin 2x) - 6(-2p\sin 2x + 2q\cos 2x) + 9(p\cos 2x + q\sin 2x)$
$\quad = 4\sin 2x$

$\Rightarrow \quad (5p - 12q)\cos 2x + (12p + 5q)\sin 2x = 4\sin 2x$

Equating coefficients of $\cos 2x$ and $\sin 2x$ gives

$$\left.\begin{array}{l} 5p - 12q = 0 \\ 12p + 5q = 4 \end{array}\right\} \quad \Rightarrow \quad p = \dfrac{48}{169}, \ q = \dfrac{20}{169}$$

$\therefore$ the particular integral is $\dfrac{1}{169}(48\cos 2x + 20\sin 2x)$

The complementary function comes from the general solution of

$$\dfrac{d^2y}{dx^2} - 6\dfrac{dy}{dx} + 9y = 0$$

The auxiliary equation is $u^2 - 6u + 9 = 0 \quad \Rightarrow \quad u = 3$ (repeated)

so the complementary function is $(A + Bx)e^{3x}$

Therefore the general solution of the given differential equation is

$y = (A + Bx)e^{3x} + \dfrac{1}{169}(48\cos 2x + 20\sin 2x)$

Exercise 3.23

Find the general solution of each differential equation.

1 $\dfrac{d^2y}{dx^2} + 3\dfrac{dy}{dx} + 2y = 5\cos x$ **2** $\dfrac{d^2x}{d\theta^2} - 3x = \cos 2\theta - 2\sin 2\theta$

- To solve differential equations of the form
$$a\frac{d^2y}{dx^2} + b\frac{dy}{dx} + cy = f(x)$$
where $f(x)$ is an exponential function

You need to know

- How to find the general solution of the differential equation
$$a\frac{d^2y}{dx^2} + b\frac{dy}{dx} + cy = 0$$

The particular integral when f(x) is an exponential function

Consider the differential equation $\dfrac{d^2y}{dx^2} + 4\dfrac{dy}{dx} + 3y = e^x$

The function on the right-hand side suggests that a function of the form $y = pe^x$ might be a solution of the differential equation. Using this as a trial solution gives

$$y = pe^x \quad \Rightarrow \quad \frac{dy}{dx} = pe^x \ \text{ and } \ \frac{d^2y}{dx^2} = pe^x$$

Substituting into the given differential equation gives $\ 8pe^x = e^x \ \Rightarrow \ p = \frac{1}{8}$

Therefore $\ y = \frac{1}{8}e^x\ $ is a solution of $\dfrac{d^2y}{dx^2} + 4\dfrac{dy}{dx} + 3y = e^x$

and the particular integral is $\frac{1}{8}e^x$

The auxiliary equation is

$$u^2 + 4u + 3 = 0 \quad \Rightarrow \quad (u + 1)(u + 3) = 0 \quad \Rightarrow \quad u = -1 \text{ or } -3$$

Therefore the complementary function is $Ae^{-3x} + Be^{-x}$

Hence the general solution of

$$\frac{d^2y}{dx^2} + 4\frac{dy}{dx} + 3y = e^x \ \text{ is } \ y = Ae^{-3x} + Be^{-x} + \frac{1}{8}e^x$$

The failure case

Now consider the differential equation $\dfrac{d^2y}{dx^2} - 4\dfrac{dy}{dx} + 3y = e^x$

The auxiliary equation is

$$u^2 - 4u + 3 = 0 \quad \Rightarrow \quad (u - 1)(u - 3) = 0 \quad \Rightarrow \quad u = 1 \text{ or } u = 3$$

so the complementary function is $Ae^{3x} + Be^x$

If we use $y = pe^x$ as a trial solution we get
$y = Ae^{3x} + Be^x + pe^x = Ae^{3x} + Ce^x$ which is only the general solution of the left-hand side of the given equation.

This means we cannot use $y = pe^x$ when the complementary function already includes a multiple of e^x.

Instead we use $y = pxe^x$ as a trial solution, giving

$$y = pxe^x \quad \Rightarrow \quad \frac{dy}{dx} = pxe^x + pe^x \ \text{ and } \ \frac{d^2y}{dx^2} = pxe^x + 2pe^x$$

Substituting into the given differential equation gives

$$p(xe^x + 2e^x) - 4p(xe^x + e^x) + 3pxe^x = e^x \quad \Rightarrow \quad p = -\frac{1}{2}$$

Therefore $y = -\frac{1}{2}xe^x$ is a solution, so the general solution of the given differential equation is

$$y = Ae^{3x} + Be^x - \frac{1}{2}xe^x$$

For any differential equation of the form $a\dfrac{d^2y}{dx^2} + b\dfrac{dy}{dx} + cy = f(x)$

where $f(x) = \beta e^{\alpha x}$, where α and β are constants, the general solution is given by $y = $ (complementary function) + (particular integral) where the complementary function is the solution of

$a\dfrac{d^2y}{dx^2} + b\dfrac{dy}{dx} + cy = 0$ and the particular integral depends on the

powers of e in the complementary function:

- use $pe^{\alpha x}$ when the complementary function does NOT contain $e^{\alpha x}$

- use $pxe^{\alpha x}$ when the complementary function DOES contain $e^{\alpha x}$

- use $px^2e^{\alpha x}$ when the complementary function contains both $e^{\alpha x}$ and $xe^{\alpha x}$

and where p can be found by differentiation and substitution into the given differential equation.

Example

Find the general solution of the differential equation

$\dfrac{d^2y}{dx^2} - 6\dfrac{dy}{dx} + 9y = e^{3x}$

First find the complementary function.

The auxiliary equation is $u^2 - 6u + 9 = 0 \implies u = 3$

Therefore the complementary function is $(A + Bx)e^{3x}$

This contains e^{3x} and xe^{3x} so we use $y = px^2e^{3x}$ as a trial solution.

$$y = px^2e^{3x} \implies \frac{dy}{dx} = 3px^2e^{3x} + 2pxe^{3x}$$

$$\implies \frac{d^2y}{dx^2} = 9px^2e^{3x} + 12pxe^{3x} + 2pe^{3x}$$

Substitution into the given equation gives

$$(9px^2e^{3x} + 12pxe^{3x} + 2pe^{3x}) - 6(3px^2e^{3x} + 2pxe^{3x}) + 9px^2e^{3x} = e^{3x}$$

$\implies p = \frac{1}{2}$

Therefore $y = \frac{1}{2}x^2e^{3x}$ is a solution of $\dfrac{d^2y}{dx^2} - 6\dfrac{dy}{dx} + 9y = e^{3x}$

So the particular integral is $\frac{1}{2}x^2e^{3x}$

and the general solution is $y = (A + Bx)e^{3x} + \frac{1}{2}x^2e^{3x}$

Exercise 3.24

Find the general solution of each differential equation.

1 $\dfrac{d^2y}{dx^2} - 2\dfrac{dy}{dx} + y = e^{2x}$ **2** $\dfrac{d^2y}{dx^2} + 2\dfrac{dy}{dx} + 5y = 4e^{-x}$

3.25 Using boundary conditions

Learning outcomes

- To summarise the general solution of differential equations of the form
$$a\frac{d^2y}{dx^2} + b\frac{dy}{dx} + cy = f(x)$$

- To find the solution of
$$a\frac{d^2y}{dx^2} + b\frac{dy}{dx} + cy = f(x)$$
given boundary conditions

You need to know

- How to differentiate standard functions and products of functions

- The meaning of the auxiliary equation

- How to find a particular integral from a trial solution

Summary of general solutions of second order linear differential equations

$$a\frac{d^2y}{dx^2} + b\frac{dy}{dx} + cy = 0$$

Auxiliary equation $au^2 + bu + c = 0$	General solution
real and distinct roots, α and β	$y = Ae^{\alpha x} + Be^{\beta x}$
repeated root, α	$y = (A + Bx)e^{\alpha x}$
complex conjugate roots, $\alpha \pm i\beta$	$y = e^{\alpha x}(A\cos\beta x + B\sin\beta x)$

$$a\frac{d^2y}{dx^2} + b\frac{dy}{dx} + cy = f(x)$$

General solution: $y = $ complementary function $+$ particular integral where the complementary function is the solution of $a\frac{d^2y}{dx^2} + b\frac{dy}{dx} + cy = f(x)$

and the particular integral comes from a trial solution that depends on the form of $f(x)$

$f(x)$	Trial solution
polynomial or constant	polynomial of same order as $f(x)$, e.g. $f(x) = 3x - 2$, $y = px + q$
trigonometric: $u\cos\alpha x$ $v\sin\alpha x$ $u\cos\alpha x + v\sin\alpha x$	$y = p\cos\alpha x + q\sin\alpha x$
exponential: $ue^{\alpha x}$	$y = pe^{\alpha x}$ when $e^{\alpha x}$ is not part of the complementary function $y = pxe^{\alpha x}$ when $e^{\alpha x}$ is part of the complementary function $y = px^2e^{\alpha x}$ when $e^{\alpha x}$ and $xe^{\alpha x}$ are part of the complementary function

When $f(x)$ is not one of the forms given in the table, you will be given a trial solution.

Boundary conditions

When we are given boundary conditions, i.e. corresponding values of x, y and possibly $\frac{dy}{dx}$, we can use these in the general solution of a differential equation to find the particular solution.

Example

Solve the equation $\dfrac{d^2v}{dt^2} + 4v = 8t$ given that $v = 0$ when $t = 0$ and when $t = \dfrac{\pi}{4}$

$\dfrac{d^2v}{dt^2} + 4v = 8t$ gives the auxiliary equation $u^2 + 4 = 0 \quad \Rightarrow \quad u = \pm\, 2i$

Therefore the complementary function is $A\cos 2t + B\sin 2t$

Using $v = pt + q$ as a trial solution gives $\dfrac{dv}{dt} = p$ and $\dfrac{d^2v}{dt^2} = 0$

Substituting into the given equation gives $4pt + 4q = 8t$ so $p = 2$ and $q = 0$

Therefore the particular integral is $2t$

$\Rightarrow \qquad\qquad\qquad v = A\cos 2t + B\sin 2t + 2t$

When $v = 0$ and $t = 0$: $\quad 0 = A$

When $v = 0$ and $t = \dfrac{\pi}{4}$: $\quad 0 = B + \dfrac{\pi}{2} \quad\Rightarrow\quad B = -\dfrac{\pi}{2} \quad\therefore\quad v = 2t - \dfrac{\pi}{2}\sin 2t$

Example

Solve the equation $\dfrac{d^2y}{dx^2} + y = 5e^x \sin x$ given that $y = 0$ and $\dfrac{dy}{dx} = 2$ when $x = 0$

Use $y = pe^x \cos x + qe^x \sin x$ as a trial solution.

The auxiliary equation is $u^2 + 1 \quad \Rightarrow \quad u = \pm i$ so the complementary function is $A\cos x + B\sin x$

$y = pe^x\cos x + qe^x\sin x \quad\Rightarrow\quad \dfrac{dy}{dx} = e^x\,(p\cos x + q\sin x) + e^x(-p\sin x + q\cos x)$

$\qquad\qquad\qquad\qquad\qquad\qquad = e^x(p + q)\cos x + e^x(-p + q)\sin x$

and $\dfrac{d^2y}{dx^2} = e^x\,(p + q)\cos x + e^x(-p + q)\sin x - e^x(p + q)\sin x + e^x(-p + q)\cos x$

$\qquad\qquad = 2qe^x\cos x - 2pe^x\sin x$

Substituting into the given equation gives $e^x((p + 2q)\cos x + (-2p + q)\sin x) = 5e^x\sin x \quad\Rightarrow\quad p = -2, q = 1$

so the particular integral is $-2e^x\cos x + e^x\sin x$

$\therefore \quad y = A\cos x + B\sin x - 2e^x\cos x + e^x\sin x$

When $x = 0, y = 0 \quad\Rightarrow\quad A = 2$, so $y = 2(1 - e^x)\cos x + (B + e^x)\sin x$

$\Rightarrow \quad \dfrac{dy}{dx} = -2(1 - e^x)\sin x - 2e^x\cos x + (B + e^x)\cos x + e^x\sin x$

When $x = 0, \dfrac{dy}{dx} = 2 \quad\Rightarrow\quad B = 3, \quad\therefore\quad y = 2(1 - e^x)\cos x + (3 + e^x)\sin x$

Exercise 3.25

1 Solve the equation $\dfrac{d^2y}{dx^2} - \dfrac{dy}{dx} - 2y = 10\sin x$ given $y = 0$ and $\dfrac{dy}{dx} = 1$ when $x = 0$

2 Solve the equation $\dfrac{d^2y}{dx^2} - 5\dfrac{dy}{dx} + 6y = (4x - 3)e^{-x}$ using

$y = (px + q)e^{-x}$ as a trial solution and given that $y = 0$ and $\dfrac{dy}{dx} = 0$ when $x = 0$

Learning outcomes

- To use substitution to reduce a differential equation to a form in which it can be solved

You need to know

- How to find the general solution of an equation of the form
$$a\frac{d^2y}{dx^2} + b\frac{dy}{dx} + cy = f(x)$$
- The chain rule
- The relationship $\frac{dy}{dx} = \frac{1}{\frac{dx}{dy}}$
- How to differentiate implicit functions
- How to use an integrating factor
- How to integrate by parts

Substitution

We have seen in Unit 1 that we can sometimes use a substitution to find y when $\frac{dy}{dx} = f(x)$

It is also sometimes possible to use a substitution to reduce a second order differential equation to a form that can be solved.

Consider the differential equation $x^2\frac{d^2y}{dx^2} + 2x\frac{dy}{dx} - 12y = 6$

We know that when the left-hand side is a second order linear equation, the solution often involves e^x, so we will try the substitution $x = e^u$

When $x = e^u$, using the chain rule gives

$$\frac{dy}{dx} = \frac{dy}{du} \times \frac{du}{dx} = \frac{dy}{du} \times \frac{1}{\frac{dx}{du}}$$

$$= \frac{dy}{du} \times \frac{1}{e^u} = \frac{1}{x} \times \frac{dy}{du}$$

$$\Rightarrow \qquad x\frac{dy}{dx} = \frac{dy}{du} \qquad\qquad [1]$$

Differentiating [1] with respect to x gives

$$x\frac{d^2y}{dx^2} + \frac{dy}{dx} = \frac{d}{dx}\left(\frac{dy}{du}\right)$$

$$= \frac{d^2y}{du^2} \times \frac{du}{dx} = \frac{d^2y}{du^2} \times \frac{1}{x}$$

$$\Rightarrow \qquad x^2\frac{d^2y}{dx^2} + x\frac{dy}{dx} = \frac{d^2y}{du^2} \qquad\qquad [2]$$

Expressing the given equation as $x^2\frac{d^2y}{dx^2} + x\frac{dy}{dx} + x\frac{dy}{dx} - 12y = 6$

we can now substitute $\frac{d^2y}{du^2}$ for $x^2\frac{d^2y}{dx^2} + x\frac{dy}{dx}$ and $\frac{dy}{du}$ for $x\frac{dy}{dx}$ giving

$$\frac{d^2y}{du^2} + \frac{dy}{du} - 12y = 6$$

The left-hand side is now linear and second order, so the equation can be solved.

Substitution can also be used to transform some second order differential equations to first order equations. This usually makes the integration easier.

Consider the equation $\frac{d^2y}{dx^2} + 2\frac{dy}{dx} = 4x$

There is no term involving y in this equation, so we can reduce it to a first order equation with the substitution $u = \frac{dy}{dx}$ so that $\frac{du}{dx} = \frac{d^2y}{dx^2}$

The given equation then becomes $\frac{du}{dx} + 2u = 4x$ which can be solved using the integrating factor $I = e^{\int 2\,dx} = e^{2x}$

Therefore $e^{2x}\frac{du}{dx} + 2ue^{2x} = 4xe^{2x}$

$$\Rightarrow \qquad ue^{2x} = \int 4xe^{2x}\,dx$$

Using integration by parts gives

$$ue^{2x} = 2xe^{2x} - \int 2e^{2x}\,dx$$
$$= 2xe^{2x} - e^{2x} + A$$
$$\Rightarrow \quad u = 2x - 1 + Ae^{-2x}$$

Substituting back for u gives another first order differential equation:

$$\frac{dy}{dx} = 2x - 1 + Ae^{-2x}$$

Integrating again gives $y = x^2 - x - \frac{1}{2}Ae^{-2x} + B$

Example

Use the substitution $u = \dfrac{dy}{dx}$ to find the general solution of the differential equation

$$\frac{d^2y}{dx^2} + 2\left(\frac{dy}{dx}\right)^2 = 0$$

$$u = \frac{dy}{dx} \quad \Rightarrow \quad \frac{du}{dx} = \frac{d^2y}{dx^2}$$

$$\therefore \quad \frac{d^2y}{dx^2} + 2\left(\frac{dy}{dx}\right)^2 = 0$$

$$\Rightarrow \quad \frac{du}{dx} + 2u^2 = 0$$

This equation can be integrated by separating the variables,

i.e. $\dfrac{1}{u^2}\dfrac{du}{dx} = -2 \quad \Rightarrow \quad -\dfrac{1}{u} = -2x + A$

$$\therefore \quad u = \frac{1}{2x + A}$$

so $\dfrac{dy}{dx} = \dfrac{1}{2x + A}$

$$\Rightarrow \quad y = \tfrac{1}{2}\ln|2x + A| + B$$

Exercise 3.26

1 Use the substitution $u = \dfrac{dy}{dx}$ to find the general solution of the

   equation $\dfrac{d^2y}{dx^2} + x\left(\dfrac{dy}{dx}\right)^2 = 0$

   Given that $y = 0$ and $\dfrac{dy}{dx} = 1$ when $x = 0$, find y in terms of x.

2 Use the substitution $x = e^u$ to show that the differential equation

   $x^2\dfrac{d^2y}{dx^2} + x\dfrac{dy}{dx} + y = 0$ can be expressed as $\dfrac{d^2y}{du^2} + y = 0$.

   Hence find the general solution of $x^2\dfrac{d^2y}{dx^2} + x\dfrac{dy}{dx} + y = 0$

Section 3 Practice questions

1 A sim card manufacturer marks each sim with a unique registration code. This code consists of one digit chosen from 1 to 6, two letters not including the letters O and I, two digits chosen from 0 to 9 inclusive and ending with two letters, again not including O and I. All digits and letters can be repeated.

The manufacturer has made 20 000 000 sim cards. How many more sim cards can be made before a new format for the codes needs to be introduced?

2 Four-digit numbers are made from the digits 1, 2, 4, 6, 7 and 9. Each digit is used only once.

(a) How many different even numbers can be made?

(b) How many different numbers can be made that are greater than 4200?

3 Three coins are tossed simultaneously. Calculate the number of ways they can land so that

(a) at least one coin lands with a head uppermost

(b) at least two coins land with a head uppermost.

4 Three cubical dice are thrown and the numbers on the uppermost faces are added to form the score.

Find the number of ways in which they can land so that the score is

(a) less than 6 **(b)** greater than 10.

5 One cubical dice is biased so that when it is thrown, it is twice as likely to show a six on its uppermost face as any other score. A second cubical dice is unbiased.

One of these dice is chosen at random and then thrown. If a six shows on the uppermost face, what is the probability that the biased dice was chosen?

6 The cards numbered 2 to 9 are withdrawn from an ordinary pack of 52 playing cards to form a smaller pack. Three cards are drawn from this smaller pack.

Calculate the probability that they all show the same number.

7 Of the 50 members of a cricket club,

27 are batsmen,

27 are bowlers,

16 are wicket keepers,

8 are batsmen and bowlers,

3 are batsmen and wicket keepers,

5 are bowlers and wicket keepers,

8 are neither batsmen nor bowlers nor wicketkeepers.

(a) Draw a Venn diagram to show this information.

(b) One member of the club is chosen at random. Calculate the probability that the person chosen is not a batsman, nor a bowler, nor a wicket keeper.

8 A bag contains 5 red discs, 8 blue discs and 6 white discs.

One disc is removed at random and not replaced, then a second disc is removed at random.

Calculate the probability that the two discs removed are the same colour.

9 $\mathbf{A} = \begin{pmatrix} 3 & 5 & -1 \\ 6 & 0 & 4 \\ 1 & -4 & 3 \end{pmatrix}$ and $\mathbf{B} = \begin{pmatrix} 1 & 0 & 4 \\ -2 & 2 & 0 \\ 3 & -1 & -2 \end{pmatrix}$

(a) Find $\mathbf{A} + 2\mathbf{B}$

(b) Determine x and y if

$$\mathbf{A} - 3\mathbf{B} = \begin{pmatrix} 0 & x+y & -13 \\ 12 & xy & 4 \\ -8 & -1 & 9 \end{pmatrix}$$

10 $\mathbf{A} = (2 \ \ 1 \ \ 4)$ and $\mathbf{B} = \begin{pmatrix} 3 \\ 0 \\ 1 \end{pmatrix}$

Show that $|\mathbf{AB}| = 10$ but $|\mathbf{BA}| = 0$

11 Given that $\mathbf{A} = \begin{pmatrix} \cos\theta & \sin\theta \\ \sin\theta & -\cos\theta \end{pmatrix}$ show that $\mathbf{A}^2 = \mathbf{I}$

12 Given $\mathbf{A} = \begin{pmatrix} 1 & x & 2 \\ 1 & 0 & 4 \\ 2 & 1 & 1 \end{pmatrix}$

(a) Find the value of x for which $|\mathbf{A}| = 0$

(b) When $x = 1$ find the value of y for which

$$\mathbf{A}\begin{pmatrix} 1 \\ -1 \\ y \end{pmatrix} = \begin{pmatrix} 6 \\ 13 \\ 4 \end{pmatrix}$$

13 Find the value of a given that

$$\begin{vmatrix} 1 & 1 & 1 \\ a & 2a & 3a \\ a^2 & a^2-1 & a^2 \end{vmatrix} = -4$$

14 $\mathbf{A} = \begin{pmatrix} 1 & -1 & 4 \\ 0 & 1 & -1 \\ 2 & 0 & 1 \end{pmatrix}$ and $\mathbf{B} = \begin{pmatrix} 0 & 1 & -1 \\ 2 & 0 & 1 \\ 1 & 2 & 0 \end{pmatrix}$

   (a) Determine $\mathbf{A}^T\mathbf{B}^T$

   (b) Show that $(\mathbf{AB})^T \neq \mathbf{A}^T\mathbf{B}^T$

15 (a) Determine which of the following sets of equations are consistent:

   (i) $2x - y + 4 = 0$
   $y = 2x - 4$

   (ii) $2x - y + 4 = 0$
   $x - y - 2 = 0$

   (iii) $2x - y + 4 = 0$
   $3y = 6x + 12$

   (b) (i) Express the set of equations that has a unique solution as a matrix equation.

   (ii) Use row reduction to solve the matrix equation.

16 Show that the matrix $\mathbf{A} = \begin{pmatrix} 1 & 2 & 2 \\ 2 & 3 & 1 \\ 4 & 2 & 1 \end{pmatrix}$

   can be reduced to $\begin{pmatrix} 1 & 2 & 2 \\ 0 & -1 & -3 \\ 0 & 0 & 11 \end{pmatrix}$

   Hence find $|\mathbf{A}|$.

17 Use row reduction to solve the equations
   $$2x - y - 2z = 2$$
   $$x + 2y - z = 6$$
   $$3x - 4y + 2z = -7$$

18 Express the system of equations
   $2x + y + 3z = 4$
   $3x - y + 2z = 2$ as a matrix equation.
   $2x + 4y + 6z = -1$
   Hence show that the system is not consistent.

19 Given $\mathbf{A} = \begin{pmatrix} 3 & 2 & -1 \\ 1 & -1 & -2 \\ 0 & 1 & 0 \end{pmatrix}$ and $\mathbf{B} = \begin{pmatrix} 4 & 3 & 2 \\ -2 & 6 & 5 \\ 1 & -3 & 1 \end{pmatrix}$

   find the matrix $\mathbf{C}$ that satisfies the equation $\mathbf{AC} = \mathbf{A}^{-1}\mathbf{B}$

20 (a) Find the general solution of the differential equation $x^2 \dfrac{dy}{dx} + 2xy = \cos x$

   (b) Find the particular solution given that when $x = \dfrac{\pi}{2}$, $y = 0$

21 (a) Find the integrating factor for solving the differential equation $\dfrac{dy}{dx} - x^2y = x^2$

   (b) Find the solution given that $y = 1$ when $x = 0$

22 Find the general solution of each differential equation.

   (a) $\dfrac{d^2y}{dx^2} - 8\dfrac{dy}{dx} + 12y = 0$

   (b) $\dfrac{d^2y}{dx^2} - 8\dfrac{dy}{dx} + 16y = 0$ (c) $\dfrac{d^2y}{dx^2} + 9y = 0$

23 For the differential equation $\dfrac{d^2y}{dx^2} - 16y = 3x - 1$ find

   (a) the particular integral

   (b) the complementary function

   (c) the general solution.

24 Given that $y = a\cos 2x + b\sin 2x$ is a particular integral of the differential equation $\dfrac{d^2y}{dx^2} + 2\dfrac{dy}{dx} + 3y = 10\cos 2x$, find

   (a) the values of the constants a and b

   (b) the general solution of the differential equation.

25 Given that $\dfrac{d^2y}{dx^2} - 4\dfrac{dy}{dx} + 4y = e^{2x}$

   (a) find the complementary function

   (b) explain why $y = ae^{2x}$ where a is a constant is not a suitable particular integral

   (c) find the particular integral and hence give the general solution of the differential equation

   (d) find the particular solution given that $y = 1$ and $\dfrac{dy}{dx} = 0$ when $x = 0$

26 Use the substitution $x = e^u$ to show that the differential equation $x^2\dfrac{d^2y}{dx^2} + 2x\dfrac{dy}{dx} - 3y = 0$

   reduces to $\dfrac{d^2y}{du^2} + \dfrac{dy}{du} - 3y = 0$

   Hence find the general solution of the differential equation $x^2\dfrac{d^2y}{dx^2} + 2x\dfrac{dy}{dx} - 3y = 0$

27 Use the substitution $u = \dfrac{dy}{dx}$ to reduce the second order differential equation $x\dfrac{d^2y}{dx^2} + \dfrac{dy}{dx} - 3x = 0$

   to a first order differential equation.

   Hence find the general solution of the equation $\dfrac{d^2y}{dx^2} + x\dfrac{dy}{dx} - 3x = 0$

Index